The CRC Press
Laser and Optical Science and Technology Series

Editor-in-Chief: Marvin J. Weber

Handbooks of Laser Science and Technology

Volume I: Lasers and Masers, Edited by Marvin J. Weber

Volume II: Gas Lasers, Edited by Marvin J. Weber

Volume III: Optical Materials, Part 1, Edited by Marvin J. Weber

Volume IV: Optical Materials, Part 2, Edited by Marvin J. Weber

Volume V: Optical Materials, Part 3, Edited by Marvin J. Weber

Supplement I: Lasers, Edited by Marvin J. Weber

Thermodynamic and Kinetic Aspects of the Vitreous State

Sergei V. Nemilov

Forthcoming Handbook Titles

Supplement II: Optical Materials, Edited by Marvin J. Weber

Handbook of Laser Wavelengths, Edited by Marvin J. Weber

Thermodynamic and Kinetic Aspects of the Vitreous State

Sergei V. Nemilov, D. Chem. Sc.
Professor
S.I. Vavilov State Optical Institute
St. Petersburg, Russia

CRC Press
Boca Raton Ann Arbor London Tokyo

Library of Congress Cataloging-in-Publication Data

Nemilov, Sergei V.
Thermodynamic and kinetic aspects of the vitreous state / Sergei Nemilov.
p. cm. -- (Weber's laser science and technology series)
Includes bibliographical references and index.
ISBN 0-8493-3782-8
1. Glass. 2. Thermodynamics. I. Title. II. Series.
TP857.N46 1994
530.4′ 13--dc20

94-16247
CIP

Direct all inquiries to CRC Press, Inc., 2000 Corporate Blvd., N.W., Boca Raton, Florida 33431.

International Standard Book Number 0-8493-3782-8
Library of Congress Card Number 94-16247
Printed in the United States of America 1 2 3 4 5 6 7 8 9 0
Printed on acid-free paper

**I dedicate this book
to the respectful memory of my Teacher
Rudolf L. Myuller**

PREFACE

The monograph represents the first attempt to present logically the major problems of the vitreous state in the framework of a generalization of thermodynamic and kinetic approaches (linear irreversible thermodynamics). The criteria for glass formation from liquid, the peculiarities of structure relaxation parameters, important relations such as the Prigogine-Defay Ratio, etc., are treated in detail, as well as model concepts specific to this set of questions. Basing on newly derived relations, values of nonequilibrium thermodynamic functions for all types of glasses are systematized, together with an analysis of the connection of the values with glass structure. Attention is also given to the problems of nonergodicity of glasses, to the low-temperature heat capacity anomalies, and to other present-day problems.

The thermodynamic essence of glass annealing and memory effects, as well as their kinetic aspects, are considered. Viscous flow theory is treated as a constituent of the kinetic description. Various versions of this theory use the values of the configurational parts of thermodynamic functions. Special attention is paid to the problem of cooperativity and its connection with structure. The analysis of the interrelations between structural and mechanical relaxations is given, based on the known laws of nonexponential behavior vs. time and temperature.

The book cites numerous results of structural, spectroscopic, and other investigations of glasses and for melts of varying natures (inorganic and organic low-molecular glasses, polymers, and crystals with properties of glasses). The results are systematized and used to illustrate the most important points of the theory. From the bibliography, the history of the various issues and questions can be followed. Actual problems having no solutions at present are also formulated.

The book is written without unnecessary formalism and most of the sophisticated theses are followed by elementary explanations. The book should be helpful to scientists and engineers working in various specializations, and especially to students and postgraduates.

TABLE OF CONTENTS

Introduction

This book is addressed to a wider range of readers than might be thought, based on its title. Its purpose is to answer the question "What is glass (or the glassy state)?" in terms of classical concepts of thermodynamics and kinetics. Such an answer pursues not only the aim of defining the notion of glass, this being treated in different ways depending on the approach and chosen criteria; but, as such, this task is not very important. Much more important in practice are the consequences of this answer, specifically: "What could and couldn't be expected of the vitreous state, and how and within what limits can glass properties be regulated according to the specific nature of glass as a certain state of substance?" The system of such concepts is of interest to anyone who deals with glass in all of the various areas of its applications (optics, spectroscopy, opto-electronics, acoustics, building, etc.) and who is interested in the nature of the state he is dealing with. On the other hand, this book should provide useful information for those theoreticians who either develop conceptions of this state of matter or are concerned with certain physical phenomena in glassforming matrices.

For the last 50 to 60 years, knowledge of the nature of the vitreous state has evolved significantly. Since the classical works by Tammann[1] in the beginning of the century, a conviction has been established that glass is a kind of supercooled liquid. At that time the only way for obtaining any glass was by the melt cooling method. However by its behavior, glass manifests itself as a solid. This always enabled one to describe glass properties in terms of solid state principles. These, however, were established first of all in the world of crystalline structures (e.g., crystallography, dynamic theory of crystal lattices, etc.). Therefore it is clear why as far back as early 1930s, there appeared a classical work by Zachariasen[2] advocating crystallochemical criteria for promoting glass formation. These criteria determine the possibility of arranging a disordered spatial structure (such as that used to understand the liquid state and, consequently, the glass structure) based on the principles of spatial connection of elementary structural polyhedra. The structure of these polyhedra is inherited from a crystal's structure and the energy of such a network differs only slightly from that of a crystalline lattice.*

* For the sake of historical fairness, the work by Frankenheim[3] should be mentioned. This author was apparently the first one to point out the similarity between the vitreous and crystalline states. He supposed that glass consisted of very small crystals of various sizes. Upon being heated, smaller crystals melt before larger ones, thus creating a sort of "lubrication" which allows all of the material to flow at temperatures much lower than the melting temperature of large crystals. This is how the process of glass softening was understood at that time. At the beginning of the 19th century, such ideas were essentially ahead of their time. In 1995 (in *Sov. J. Glass Phys. Chem.*, vol. 21, Interperiodica Inc. Ed.) a paper will be published by S.V. Nemilov entitled "Moritz Ludwig Frankenheim as the founder of crystallite glass-structure hypothesis" devoted to the origin of the first main glass-structure concepts in the beginning of the nineteenth century.

Subsequently hundreds of papers appeared which developed structural and kinetic ideas of the origin and nature of the vitreous state. The totality of this human thinking exhibited a diversity of approaches, ideas, models, and experimental results. It is not appropriate to go into details here. A certain part of those works answer the question: "Why does a glass have exactly this but not another chemical composition, and what are the criteria needed for that?" Such models use the crystallochemical criteria proper, as well as parameters determining the nature of chemical bonds in a system (different degree of ionicity or covalency, etc.), correlations between chemical bond strength and melting temperatures of crystals, concentrations of electrons per atom, and so on. Almost all of these works are speculative. Another part answers the question "How and under what conditions can glass be obtained?" Here a kinetic factor, namely, the melt cooling rate is used. In connection with the theory of nucleation and growth of crystallization centers, this explanation is given for the greater or lesser tendency of a glass melt to crystallize, i.e., to the melt stability with respect to the crystallization process. An equivalent but separate branch in this direction is the kinetic theory of glass formation from liquid having avoided crystallization, as a special state (glass transition theory). Certain fundamentals of this concept were laid by Tool.[4] This theory will be treated in this book in the essence of its classical version within the framework of nonequilibrium thermodynamics. In this context, the theory of kinetic coefficients including viscosity also pertains.

The third significant part of the works devoted to the nature of the vitreous state answer the question: "What structure does the glass have when this state has been obtained?" These works first of all are comprised of results of diffraction investigations of medium- and large-angle scattering (X-rays and neutron diffraction), then studies on glass structure using various resonance methods (first of all, by the NMR method) and, to some degree, vibrational spectroscopy methods. They give answers to questions about the type (nomenclature) and structure of elementary polyhedra of glass networks (short-range order), about the nature of their spatial linking (middle-range order), and about different kinds of spatial ordering. Small-angle scattering studies answer the question about the inhomogeneous structure of glass as a consequence of liquid structure fluctuations or as a consequence of existing stable inhomogeneities (a sort of liquid phase separation inhomogeneities). This group of studies uses strict physical methods. To obtain structural information, well-developed mathematical treatment methods are used. However, this group of works is not free of uncertainty either, because the final structural picture may depend on a number of natural simplifications and assumptions in the methods of treating the data. Nevertheless, these results on glass structure are still most valid and we shall use them. To achieve the greatest correspondence between models and real structural and thermodynamic data, computer simulation methods are used (Monte-Carlo and molecular dynamics methods). The book by Brawer[5] is already devoted to this problem.

The suggested classification is rather conventional and incomplete. Combined and complex approaches are also possible. There are several modern books devoted to the vitreous state (e.g., Feltz,[6] Zarzycki[7]). Each book, with more or less thoroughness, deals with the entire complex of the problems mentioned above. In these books the reader will also find comprehensive characteristics of modern glass manufacturing methods, various fields of glass applications, and the peculiarity of the complex of glass and melt properties. But almost none of them elucidates the vitreous state problem in terms of a unified concept, except, possibly, the book by Jones.[8] This author's approach is the nearest one to the idea of this book. The choice of a unified system leads inevitably to omitting some information, both general and particular. Therefore, I hope that the reader will have gotten acquainted with such information elsewhere, perhaps by practical experience.

The basis of this book is the thermodynamic approach. Its advantage, besides its strictness, is that thermodynamics disengages itself from the real structure of matter. Thermodynamic results are always obtained in general form. From this point of view the formation of glass from a liquid can be described in a universal way. The consequences of this theory should therefore be general. However, nonequilibrium thermodynamics includes kinetic coefficients, too. Therefore, introducing these coefficients by concrete molecular-kinetic models allows general relations to be specified and helps to understand the specific features of the vitreous state concerning the real structure.

In this book the author proceeds to some extent from his own scientific interests. But the text attempts to represent all of the most important results that have been obtained in the world and thus not only the author's publications.

The specific character of general consideration demands involving facts on all types of vitreous substances, i.e., inorganic and molecular organic glasses and polymers. It is natural that I had to confine myself to those aspects of the investigations which are of most interest. As this is the first attempt at such generalization, I have tried to summarize not only the most important concepts and directions, but also to identify and formulate urgent and unsolved problems.

The book is arranged in such a way that the reader can skip special sections in which concrete thermodynamics and kinetics treatment methods are used without loss of essential comprehension. The most important results from such strict consideration can be easily understood from the text supplied.

I wish to thank Professor Leonid B. Glebov for supporting my efforts at writing this book, and Ms. Alexandra E. Yakuninskaya, who translated the text from Russian into English and eliminated a number of essential defects.

Also, I wish to thank Drs. Natalia V. Borisova and Leonid V. Maksimov who have reviewed several sections of the book and provided useful remarks.

Sergei V. Nemilov
St. Petersburg, Russia

I Origin of Glass: Thermodynamic Approximation

A. THERMODYNAMIC FUNCTIONS AND THERMODYNAMIC VARIABLES

To understand the nature of the vitreous state, it is essential to define thermodynamically the state of a liquid that gives rise to glass. For this purpose, the most essential thermodynamic definitions need to be introduced. This system of definitions goes back to the founder of chemical thermodynamics, J. W. Gibbs, and to the works by H. Helmholtz, M. Planck, W. Nernst, and others.

A thermodynamic system is a part of the physical world, somehow separated from the exterior, that can do work on its surroundings or can be subject to work done by those surroundings. A thermodynamic system is capable of exchanging heat with its surroundings. There exist closed and open systems. A closed system has no mass exchange with its surroundings whereas an open one does have such an exchange. One kind of closed system comprises isolated systems having neither matter nor energy exchange with the exterior. Adiabatically isolated systems can change only by being subjected to some work (e.g., fast gas compression within impenetrable heat boundaries). Isothermal systems exchange heat (Q) with their exterior so that the temperature inside and outside such a system is the same. The notion of "heat" is not identical to that of "energy." A system is characterized by the following functions: the internal energy U, the heat content H, the entropy S, and the Gibbs free energy G or the Helmholtz free energy F. The free energy determines the ability of a system to do the maximum possible useful work W.

For any variation in the states of systems, one only deals with changes in the above functions. Their absolute values remain unknown. The choice of the reference state introduces the starting point for calculations. For instance, as a reference state one can choose the state of a crystal at absolute zero. Values of thermodynamic functions corresponding to the states of that crystal, liquid, and vapor at high temperatures will start from that state. For that (reference) state the values of the above functions are zero.

The parameters of state are those variables the values of which can be specified. There is an individual set of such parameters for each type of system. These are, for example, the temperature T and the pressure p. If these are constant, the Gibbs isobaric-isothermal thermodynamic potential is considered. If the geometrical volume V and the temperature are kept constant, the Helmholtz isochoric-isothermal potential F is considered. Relations between the U, H, G, F, and S functions and variable parameters of state are regulated for reversible processes by the First and Second Laws of Thermodynamics and Maxwell's equations. Here we list only those that we will need later:

$$dU = dQ - dW \quad \text{(The First Law)} \tag{1}$$

In the simplest mechanical case dW = pdV and

$$dS = dQ / T \quad \text{(The Second Law)} \tag{2}$$

Some consequences of the first and second laws are the following:

$$dU = TdS - pdV \tag{3}$$

$$G = H - TS \tag{4}$$

$$F = U - TS \tag{5}$$

$$\left(\frac{dG}{dT}\right)_p = -S \tag{6}$$

$$\left(\frac{dG}{dp}\right)_T = V \tag{7}$$

$$\left(\frac{dH}{dS}\right)_p = T \tag{8}$$

$$\left(\frac{dF}{dT}\right)_V = -S \tag{9}$$

$$\left(\frac{dF}{dV}\right)_T = -p \tag{10}$$

$$\left(\frac{dU}{dS}\right)_V = T \tag{11}$$

Evidently, the entropy S at T and p = constant (or at T and V = constant) determines that part of the total heat content H (or of the internal energy) of the system that cannot transform into useful work. Various states of the system (at T, p = constant with changing volume and at T, and V = constant with changing pressure are connected with the performance of thermodynamic processes. Reversible processes are defined as those for which the transition from the initial state to the final one proceeds without waste of heat Q (i.e., without friction). In particular, for isothermal heat transfer processes

$$\frac{dQ}{T} = dS \tag{12}$$

For irreversible processes

$$\frac{dQ}{T} < dS \tag{13}$$

and the exterior energy increases because of the production of waste (useless) work.

For systems in equilibrium the notion of "functions of state" is introduced. These are U, H, G, F, V, and S. Their differentials are expressed by the total differentials in the state variables, only in this case the system is uniquely defined and its internal structure is an exclusive function solely of the specified parameters of the state. For example, a liquid can change its structure with gradually changing temperature. It passes through a sequence of equilibrium states connected by the condition of reversibility, but its internal energy, thermodynamic potential, entropy, and volume are always defined as single-valued functions and, consequently, its structure is also uniquely defined. If this condition is not satisfied, U, H, G, etc., are not termed "functions of state."

A system may consist of different parts; e.g., liquid may have a vapor (gas phase) above it. A melt may contain a crystalline phase. "Phase" means a compositionally homogeneous part or complex of parts of a heterogeneous system having the same temperature and thermodynamic properties at every point. Phases of an equilibrium system are in equilibrium to each other. This means that the chemical potentials of the components (i.e., chemical individuals of the ith type forming the system), defined as $(\partial G/\partial m_i)_j$ (m_i is the mass of the ith component in the jth phase), are equal to one another for an actual complex of existing phases.

Liquid coexists in equilibrium with its vapor, crystal with its liquid, and so on. The conditions of a system being in equilibrium restrict the variety of phase equilibria and the number of these phases (the Gibbs Phase Rule is not to be considered here).

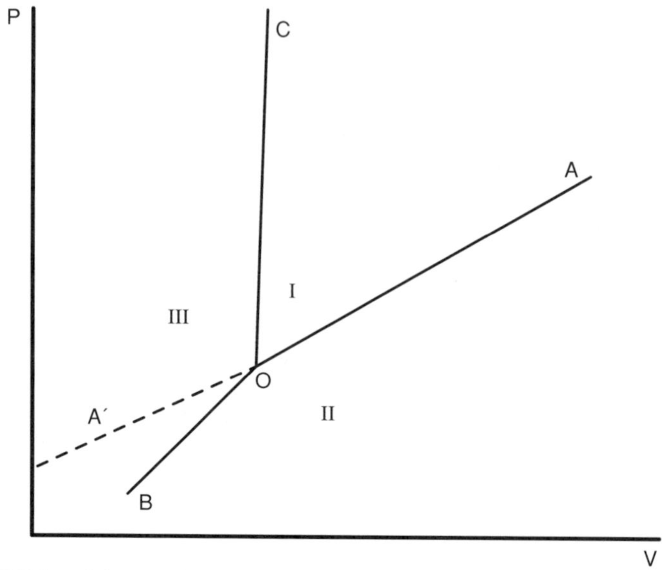

FIGURE 1. Scheme of phase equilibria in a one-component system. I, area of liquid; II, area of vapor; III, area of crystal.

The notion of "phase" in thermodynamics is more generalized than those of "solid" and "liquid" in mechanics. For every substance there is an area of stable states in the coordinates of thermodynamic variables. This is shown by the scheme in Figure 1. In area I a liquid always exists in equilibrium either with its vapor only (area II), or together with another phase (area III) at lower temperatures or at higher pressures. Transition from one phase to another is a phase transition. The lines OA, OC, and OB are the lines of equilibrium phase transitions I $\leftrightarrow$ II, I $\leftrightarrow$ III, III $\leftrightarrow$ II. Within field I, liquid is absolutely stable. The dotted line OA′ indicates the area of possible metastable states of liquid (supercooled liquid). The state of vapor II in this area is also metastable.

Nature has arranged that the "other" phase III into which liquid transforms upon cooling is exactly the crystal phase. The internal structure of this phase is not defined by thermodynamics. In mechanics the notion of a "liquid" is associated with noticeable fluidity. "Solid" does not possess the property of fluidity as defined in mechanics. Criteria of this type do not work in thermodynamics. The viscosity of a liquid may be infinitely high (up to ~ 10^{13} Poise, P, and more) but this state is still a liquid if its area corresponds to the regions shown by Figure 1.

In the first half of the 20th century Th. De Donder[9] developed a rigorous method for the thermodynamic description of systems that were not in the state of internal equilibrium which is very convenient for practical applications. That method was first widely used in chemistry.[10] But the capabilities of the method were much wider and its ideas were used when extended to the thermodynamic theory of relaxation. Here we dwell briefly on the basic ideas of the method within the limits of closed systems, which are most important for us.

Consider a one-phase system comprised of a mixture of 1, ..., j substances capable of chemical interaction.[10] The latter results in j + 1, ..., c products of the reaction. If the corresponding stoichiometrical coefficients are $\nu_1, ..., \nu_c$, the reaction equation is

$$\nu_1 R_1 + \nu_2 R_2 + \cdots + \nu_j R_j \rightarrow \nu_{j+1} R_{j+1} + \cdots + \nu_c R_c \tag{14}$$

The mass variation of each substance with respect to its initial mass in the starting mixture (labeled by a superscript o) at every instant is expressed by the coefficient ξ. It is named "the reaction extent" or "the reaction coordinate":

$$\left.\begin{aligned} m_1 - m_1^o &= \nu_1 M_1 \xi \\ m_i - m_i^o &= \nu_i M_i \xi \\ m_c - m_c^o &= \nu_c M_c \xi \end{aligned}\right\} \tag{15}$$

where $M_1, ..., M_c$ are molecular weights of the substances $R_1, ..., R_c$. In the initial state $\xi = 0$. At the completion of the reaction running from the left to the right $\xi = 1$. Differentiating (15), we have

$$\frac{dm_1}{\nu_1 M_1} = \frac{dm_2}{\nu_2 M_2} = \cdots = \frac{dm_c}{\nu_c M_c} = d\xi \tag{16}$$

The reaction rate is defined as

$$v = \frac{d\xi(t)}{dt} \tag{17}$$

Evidently, $d\xi/dt = v(T,p,\xi)$. The internal energy of such a system and the other thermodynamic potentials are unique functions not only of T, p, T, and V, but of ξ as well:

$$\begin{aligned} H &= H(T, p, \xi) \\ G &= G(T, p, \xi) \\ U &= U(T, V, \xi) \\ F &= F(T, V, \xi) \end{aligned} \tag{18}$$

Thus, the reaction coordinate is a proper thermodynamic variable, along T, p, V, etc. Increments of those functions are sums of the increments that are required by classical thermodynamics assuming the constancy of composition (ξ = constant) and of the increments caused by changing the system composition. For example,

$$dH = \left(\frac{\partial H}{\partial T}\right)_{p,\xi} dT + \left(\frac{\partial H}{\partial p}\right)_{T,\xi} dp + \left(\frac{\partial H}{\partial \xi}\right)_{T,p} d\xi \tag{19}$$

and

$$dU = \left(\frac{\partial U}{\partial T}\right)_{V,\xi} dT + \left(\frac{\partial U}{\partial V}\right)_{T,\xi} dV + \left(\frac{\partial U}{\partial \xi}\right)_{T,V} d\xi \tag{20}$$

Summarizing the First and Second Laws, [(1), (2)] takes the form

$$dU = TdS - pdV - dQ' \tag{21}$$

where dQ' is the so-called "uncompensated heat" connected with irreversible chemical conversion inside the system.* At constant T and V

$$dQ' = -dF > O \tag{22}$$

and at constant T and p

$$dQ = -dG > O \tag{23}$$

where dF and dG are the changes of the free energy (Helmholtz and Gibbs, respectively) due to chemical reaction proceeding.

Chemical affinity A denotes the function defined by the relation

$$dQ' = Ad\xi \geq O \tag{24}$$

The sign ">" holds for the case when the process rate is nonzero, the sign "=" is for the case of reversible equilibrium (the equilibrium system).

The rate of a chemical reaction and the affinity are bound up by De Donder's inequality:

$$\frac{dQ'}{dt} = A\frac{d\xi}{dt} = Av \geq O \tag{25}$$

The affinity is defined as the free energy change (with the sign "–") corresponding to the unit reaction run:

$$-A = \left(\frac{\partial H}{\partial \xi}\right)_{T,p} - T\left(\frac{\partial S}{\partial \xi}\right)_{T,p} = \left(\frac{\partial G}{\partial \xi}\right)_{T,p} \tag{26}$$

* In the equation for the Second Law of Thermodynamics (2), dQ is the amount of heat exchanged by the system with its exterior; dQ and dQ' therefore have different physical meanings.

or

$$-A = \left(\frac{\partial U}{\partial \xi}\right)_{T,V} - T\left(\frac{\partial S}{\partial \xi}\right)_{T,V} = \left(\frac{\partial F}{\partial \xi}\right)_{T,V} \tag{27}$$

The meaning of the second derivatives of the thermodynamic potential (thermodynamic coefficients) can also be defined similarly. We shall return to this problem later in connection with their specific applications.

It is important to note that De Donder's method of nonequilibrium thermodynamics does not introduce any new thermodynamic principles. It is just a method for describing nonequilibrium systems. Its results correspond completely to the First and Second Laws of Thermodynamics. This method is described in detail by Prigogine and Defay.[10]

B. CLASSIFICATION OF STATES ACCORDING TO STABILITY FEATURES

The concept of stability is extremely important for our discussion. One should distinguish the conditions of stability for adiabatically isolated systems from those for isothermal systems. Systems of the first type are stable when their entropy reaches its maximum. Systems of the second type achieves stability when their thermodynamic potential reaches its minimum. The latter type is of primary importance in our considerations.

Generally speaking, there can be three types of equilibrium in mechanics as well as in thermodynamics. Thus, a parallelepiped laid on a plane on its largest side has the absolute minimum potential energy. It is the most stable state of the system. The same parallelepiped lying on the plane on one of its minor sides has a relative minimum of its potential energy. When it is tilted slightly nothing happens, but if it is tilted more, the parallelepiped can overturn on its largest side. In this process, the potential energy overcomes a certain barrier. The energy of the body in its new position achieves the absolute minimum. In this case the initial equilibrium state is "metastable." Thermodynamic fluctuations or small variations of the parameters of state are analogous to tilting of the parallelepiped. The metastable state is stable with respect to small, but unstable with respect to large, fluctuations or influences. For example, liquids at temperatures below the melting point that have no crystallization nuclei can remain in this state of metastable equilibrium for an infinitely long time. Small fluctuations do not disturb this equilibrium, but introducing any heterogeneous nucleus of crystallization (which can be, e.g., on a phase separation surface) creates significant local fluctuations and the liquid gradually becomes crystallized.

The third case is related to an absolutely unstable equilibrium (labile state). It corresponds to a parallelepiped accidentally standing on one of its vertices without falling or to a cone standing similarly on its vertex. The center of gravity lies on the vertical line passing through the supporting point. Any infinitesimal displacement of the center of gravity will cause the body to fall. An analogous thermodynamically labile system is unstable to any infinitesimal fluctuation. (A variation of the latter type of system includes frozen unstable systems. This is the same situation, but all the objects are in a medium of extremely high viscosity.) Obviously, small fluctuations would not cause the body to fall if the time of observation was short enough. Similarly, a long time is required or a strong influence on a body is necessary to overcome the resistance of a fluid medium. Here one needs high precision to observe displacements. As we shall see later, all glasses belong to these types of systems. Surprisingly, even in recent literature devoted to the vitreous state the above notions of stability are often confused.

The preceding concepts are the minimum necessary for the discussions that follow. A most brief, but strict and profound account of these concepts can also be found in the book by Münster.[11]

C. GENESIS OF NONEQUILIBRIUM STATES OF LIQUID

The thermodynamic theory of the glass transition has evolved from the thermodynamic theory of relaxation in liquids, the foundation of which was laid by Mandel'shtam and Leontovich.[12] This theory, in the case of acoustic relaxation, describes the response of a system to a periodic (sinusoidal) action of frequency ω. If the system is defined by the external state parameters (T,p), (T,V), etc., the periodic variation of at least one of them (e.g., pressure) produces areas of disturbance in the liquid (e.g., compression or rarefaction). These areas either decay or persist intact, depending on the relation of the structural relaxation time (τ) to the period of the variable variation ($2\pi\omega^{-1}$). At sufficiently high frequencies ω, the liquid behaves as a solid and is characterized by solid-like coefficients (these are experimentally measurable second derivatives of a thermodynamic potential: compressibility, heat capacity, etc.). The operative moduli (e.g., bulk modulus K_T, heat capacity C_V, etc.) are the sums of a limiting instantaneous part (labeled ∞) and a relaxational part (labeled r)

$$\tilde{K}_T = K_{T\infty} - \frac{K_{Tr}}{1+\tau\, d/dt} \quad \text{and} \quad \tilde{C}_V = C_{V\infty} + \frac{C_{Vr}}{1+\tau\, d/dt} \tag{28}$$

where $d/dt \sim i\omega$ (Solov'ev[13]). The quantities determined by the relaxation parts are nonzero at low frequencies ($\omega \to 0$). Then the liquid properties are indistinguishable from those of stable liquid (i.e., K_{T0}, C_{V0}, etc.).

The theory tells us that the cause of the solid-like state is the freezing of the internal structural parameter ξ, the variation of which can be normalized to

1 and the physical meaning of which is not defined beforehand. This may be the degree of deviation of atomic (molecular) correlation functions from their equilibrium value (structural nonequilibrium), the extent/degree of chemical conversion or of transformations of one isomer into another (chemical nonequilibrium), the degree of packing density, the degree of molecular orientation, etc. As a function, the parameter ξ is completely similar to the reaction coordinate in the De Donder's thermodynamic method.[9] However, it can acquire wider physical meaning.

In an equilibrium liquid (at low frequencies), ξ is related uniquely to the variables (T,p), (T,V), etc. Therefore, the description of thermodynamic potential variations (in terms of classical thermodynamics) does not need to involve the notion of a structural parameter.

In a nonequilibrium liquid ξ has no time to follow the variations of those parameters, so there appears an excess of a thermodynamic potential with respect to equilibrium. That excess is defined by the nonequilibrium character of ξ. The description of time variations of such a system as it approaches the equilibrium state is provided by thermodynamic relaxation theory.[13]

Suppose that another external parameter, namely the temperature T of the system, varies and that we confine ourselves to a part (e.g., $\lesssim$ 1/4) of this variational period when the temperature is falling. This procedure is actually carried out by an experimenter when he cools a melt to obtain a glass. At low cooling rates, the current values of the elastic modulus are the same as its equilibrium value K_{T0}. In this case the variations of ξ determined by the relaxation time τ are able to follow the temperature variations. Irrespective of the form of the assumed law (linear, sinusoidal, etc.) by which T varies, we inevitably have a frozen state (glass) below a certain temperature T_g. This is due to the fact that, as the temperature goes down, the viscosity increases and the structural relaxation time τ becomes too long for the structure to relax ($\omega\tau/2\pi \rightarrow \infty$). The abruptness of T_g appearing as a break point in the property vs. temperature curves depends on the instantaneous rate of change in the state parameter in that region of the system states. As this happens, the ξ value freezes and the property (e.g., the elastic modulus) falls into the region of solid-like or instantaneous values ($K_{T\infty}$). The only condition for the glass transition is that the rate of the parameter variation must be substantially higher than that of structural relaxation. We then have a nonequilibrium state.

To simplify the thermodynamic tools, the thermodynamic glass transition theory assumes an artificial situation, i.e., that above T_g the equilibrium structure always exists, whereas at T_g the external parameter of state (temperature, pressure) changes with an infinitely high rate. That is why the range of the glass transition range gets reduced to the point T_g. Similar reasoning would apply to a change in any parameter of state (when pressure varies, p_g appears). Likewise, a nonequilibrium state may arise as a result of infinitely rapid heating, expansion, etc.

When choosing different values of T_g (different relaxation times τ), the same operation of a glass transition can be carried out, in principle, at each of them. Therefore the thermodynamic glass transition theory describes the formation of glass at any viscosity.

The question of how long such a glass would exist depends upon scale of the observation time. At small values of the relaxation time (at low viscosities), e.g., $\tau \sim 10^{-4}$ s^{-1}, the period of observation of the state should be less than that value. It is natural that one can examine such a state only for times as long as $\sim 10^2$ s and more. This condition includes temperatures below T_g where the T_g values must correspond to viscosities of 10^{13} P and more.

Returning to the possibility of a gradual lowering of the temperature, we can say that in terms of thermodynamic relaxation theory, the glass transition will occur at different T_g values (or in different ranges including T_g) depending on the cooling rate selected. It is quite natural to suppose that a uniform criterion of $\omega\tau$ = constant corresponds to the glass transition condition, as well as the fact that the cooling rate q and ω are correlated linearly in this temperature range. If the relaxation time (and viscosity) grows exponentially with decreasing temperature in the same temperature range (this is a natural assumption, too), then the dependence of T_g on the cooling rate q, as follows from the thermodynamic relaxation theory, must be $T_g = A[\ln(B/q)]^{-1}$ (where A and B are constants). This relationship corresponds to experimental results.

The thermodynamic theory of the glass transition applied to problems connected directly with thermodynamics is removed from the specific regime of the most nonequilibrium state genesis (i.e., of the glass genesis). However, it assumes in general any regime, and in terms of special periodic changes in external parameters provides the correct dependence of the temperature (or pressure) of the transition from liquid to glass vs. the rate of parameter change.

D. THERMODYNAMIC MODEL OF THE VITREOUS STATE GENESIS

The results presented in this and following sections were in general formulated by Davies and Jones.[14] Later the same results were obtained by other authors[15-18] by slightly different methods. To create the most general picture of the phenomenon, we shall formulate the thermodynamic glass transition theory in terms of criteria of thermodynamic stability, as has already been done in our previous works (Nemilov[19,20]). This seems to be the most explicit way because it exhibits the maximum complete set of thermodynamic and kinetic criteria for the glass transition. The methods of thermodynamic relaxation theory, described in most detail in the review by Solov'ev,[13] are used below. Solov'ev has generalized the Mandel'shtam and Leontovich[12] theory of acoustic relaxation in liquids which actually used De Donder's method of nonequilibrium thermodynamics.[9]

Classical chemical thermodynamics, according to Gibbs, applies the notion of "stability criteria" in connection with the possibility of a new phase arising within the existing phase with respect to the adjacent states. Adjacent states differ infinitesimally from the state under consideration by their temperature, pressure, or composition.[21] Stability criteria are given by the system of inequalities:

$$(dT/dS)_a > 0 \quad (dp/dV)_a < 0 \quad (d\mu_i/dm_i)_a > 0 \tag{29}$$

where the label *a* indicates the method of fixation of parameters; μ_i and m_i are respectively the chemical potential and mass of the ith component, and T, S, p, and V are the temperature, entropy, pressure, and volume of the system. The initial state may correspond to a state of stable or metastable equilibrium. The meaning of the inequalities in (29) is very simple. With increasing temperature the entropy of a stable system increases and its heat capacity cannot be negative. With increasing pressure, the volume of such a system is reduced, and with the growth of any component associated with one of the phases of the system, the chemical potential of this phase increases.

A labile, i.e., absolutely unstable, state corresponds to the criteria

$$(\Delta S)_{U,V,m_1,\cdots,m_n} > 0 \quad (\Delta U)_{S,V,m_1,\cdots,m_n} < 0 \tag{30}$$

according to which the resulting system entropy always increases and the internal energy decreases (at constant entropy, volume, and component masses).

In classical thermodynamics, the time factor is absent and the way in which an unstable state becomes possible is not revealed. Nonequilibrium thermodynamics removes this deficiency by introducing the time factor into the description of a system.

Our final goal consists of defining the criteria for the genesis of thermodynamically unstable states of a homogeneous system lying adjacent to its stable states. Since a system remains homogeneous for all of the state parameters considered, i.e., no new phases arise, the latter of stability conditions (29) so far is not to be considered.

1. Basic Definitions and Assumptions of the Model

a. Parameters of State

In addition to specified external parameters of the state of the system, there exists an internal structural parameter or several such parameters. Here we confine ourselves to a single parameter. Its value may be normalized and can vary from 0 to 1; it is designated as ξ. The meaning of this parameter has been already explained.

b. Vitreous State

The vitreous state is defined as a particular case where ξ = constant. Then for the specified set of parameters T,p (T,V), the glass transition point for given ξ is determined as (T_g,p_g) or (T_g,V_g)* (hereafter we shall confine ourselves to the first pair of variables). In the neighborhood of that point an equilibrium state exists (ξ is in equilibrium); a nonequilibrium state may also exist for which ξ = constant. The notion of "neighborhood of the point" implies small changes of the parameters $\delta T = T - T_r$, $\delta p = p - p_r$, and $\delta\xi = \xi - \xi_r$, where the label r in a particular case coincides with the label g. The term "nonequilibrium state" means a state that has higher thermodynamic potential than the equilibrium state. Note again, that the ξ = constant condition is introduced only for convenience. We are not interested in the future of the nonequilibrium state, whether it will persist for an infinitely long time or change at later times and the system approach the equilibrium state.

c. Possibility of a Complete Thermodynamic Description of the State under Consideration

In Reference 13 a detailed basis is given for the fact that thermodynamic description is applicable both for the ξ = constant case and for the equilibrium state case. While the latter assertion is trivial, the meaning of the former needs to be explained. If the structure of a system does not change with time and ξ is frozen, the thermodynamic potentials, entropy, volume, etc., are not defined solely by the state parameters (T, p, etc.) but also depend on the frozen structure, i.e., on the way those states have been obtained. The latter may be different. However, if the structure does not change with time (ξ = constant), the increments of the thermodynamic potentials at T,p or T,V variation are defined uniquely. Such a body can reversibly acquire and return heat to its surroundings without changing its internal structure. All changes of state will obey the relations of classical thermodynamics (1)–(11).

An example of this is a glass thermometer. Actually, all the thermodynamic investigations of the 19th century were carried out using this tool. However, it is known that over the years a reduction of the length of the glass capillaries will occur because of the material's relaxation. Until it was realized that this relaxation resulted in errors in the temperature measurements, the assumption of reversible changes of state was readily accepted. It corresponds to the actual constancy of the structural parameter ξ. For the constancy of ξ, the period of the system observation (τ') and the structural relaxation time τ must be connected by the inequality $\tau' \ll \tau$.

If the parameter ξ is treated equally with the other variables (T, p, etc.) and can vary, the thermodynamic functions become the functions of state. Their increments are expressed as the total differential in all of the variables including ξ [(19) and (20)]. Under these conditions the duration of the system observation is comparable with the structure relaxation time. The system can

* For simplicity we exclude from the number of parameters such as tensors of shear stresses, strains, electric and magnetic intensities, etc.[22]

change with time due to certain processes proceeding inside it. Its description can be complete in terms of nonequilibrium thermodynamics.

Taking these considerations into account, we will regard the Gibbs free energy as corresponding to the possibility of a complete thermodynamic description of the system both for the equilibrium state when $\xi = \xi(T,p)$,

$$G = G(T, p) \tag{31}$$

and for the absence of equilibrium when

$$G = G(T, p, \xi) \tag{32}$$

d. Properties of the Thermodynamic Functions of the States

All thermodynamic functions of each state are presumed to be smooth; i.e., they are uninterrupted, they are single-valued, and their derivatives including the second order are not infinite.

We now introduce following designations:

$$G_T = (\partial G / \partial T)_{p,\xi} \tag{33}$$

$$G_{TT} = (\partial^2 G / \partial T^2)_{p,\xi} \tag{34}$$

etc., including

$$G_\xi = (\partial G / \partial \xi)_{T,p} = -A \tag{35}$$

where A is the affinity of a structural change process, and

$$G_T = -S \tag{36}$$

$$G_p = V \tag{37}$$

are respectively the entropy and the volume of the system.

$$G_{\xi\xi} = (\partial^2 G / \partial \xi^2)_{T,p} \tag{38}$$

is the change of affinity per unit run,

$$-TG_{TT} = T(\partial S / \partial T)_p = C_p \tag{39}$$

is the isobaric heat capacity of the system,

$$-G_{pp} / V = -(\partial V / \partial p)_T / V = \kappa_T \tag{40}$$

is the isothermal compressibility of the system,

$$G_{Tp} / V = (\partial V / \partial T)_p / V = \alpha \tag{41}$$

is the thermal expansion coefficient of the system,

$$G_{\xi p} = (\partial V / \partial \xi)_{T,p} = -\Delta V \tag{42}$$

is the volume change at transformation, and

$$-TG_{\xi T} = T(\partial S / \partial \xi)_{p,T} = \Delta Q' \tag{43}$$

is the transformation heat. Thus, the first and second derivatives of G are respectively functions of state and properties of the system that are measured independently. The meaning of the above derivatives is clear from general thermodynamic relations.

e. Kinetic Assumptions

The theory[13] treats those nonequilibrium states that differ only slightly from the equilibrium one. In particular, it is considered that the approach of the system to equilibrium with changing ξ proceeds under the following conditions:

1. ξ changes only due to the affinity A differing from zero.
2. The relaxation equation is linear.
3. Relaxation has no inertia; i.e., A determines the rate (but not acceleration) of ξ's changing. The process rate is proportional to its driving force, i.e., to its affinity.

These restrictions do not depend on the causes forcing the system out of equilibrium (either temperature or mechanical influences).

These assumptions are identical in both the thermodynamic theory of the glass transition and the thermodynamic theory of acoustic relaxation.

Conditions (1) to (3) correspond to the equation

$$d\xi / dt = -\gamma G_\xi \tag{44}$$

Expressing the affinity (near the equilibrium when $G_\xi = 0$) as

$$G_\xi = G_{\xi T}\delta T + G_{\xi p}\delta p + G_{\xi\xi}\delta\xi \tag{45}$$

we have

$$\frac{d\xi}{dt} = -\gamma G_{\xi\xi}\left(\delta\xi + \frac{G_{\xi p}\delta p + G_{\xi T}\delta T}{G_{\xi\xi}}\right) = -\frac{\left(\delta\xi - \delta\xi^{e}\right)}{\tau} = -\frac{\left(\xi - \xi^{e}\right)}{\tau} \quad (46)$$

Here ξ^e is the value of ξ in the equilibrium, τ is the relaxation time, and γ is the kinetic coefficient, i.e., the transition modulus connecting τ with the coefficient $G_{\xi\xi}$

$$\tau = \left(\gamma G_{\xi\xi}\right)^{-1} \quad (47)$$

where $\gamma > 0$ because no system is free from its inner friction. The coefficient $G_{\xi\xi}$ characterizes the equilibrium state and each (T,p) point is a unique function of the specified parameters: $G_{\xi\xi} = f(T,p)$.

2. Thermodynamics of the Glass Transition

In the general case, for any state close to the specified one, i.e., in the neighborhood of the point (T,p), the free energy of the system is, using a Taylor's series expansion and confining ourselves to square terms,[13]

$$G = G^{o} + G_{p}\delta p + G_{T}\delta T + G_{\xi}\delta\xi + (1/2)(G_{pp}\delta p^{2} + 2G_{Tp}\delta p\delta T + \\ + G_{TT}\delta T^{2} + G_{\xi\xi}\delta\xi^{2} + 2G_{\xi p}\delta\xi\delta p + 2G_{\xi T}\delta\xi\delta T) \quad (48)$$

Suppose the chosen T and p are the temperature and pressure of glass transition, T_g and p_g, respectively. Then from (48) we get expressions for the equilibrium liquid free energy (hereafter the label A = 0) and for the glass free energy (hereafter the label $d\xi = 0$) in the neighborhood of the point (T_g,p_g). From the above definitions it follows that in every point of that area, except the point (T_g,p_g),

$$\Delta G = G_{d\xi=0} - G_{A=0} > 0 \quad (49)$$

and at the glass transition point the glass and the liquid are indistinguishable. It corresponds to the equality $\Delta G = 0$ (i.e., the instantaneous cooling by δT or changing pressure by δp may produce glass, and since $\delta T \rightarrow 0$ and $\delta p \rightarrow 0$, $\Delta G \rightarrow 0$ and the neighboring area reduced to a point). Similarly, at the glass transition point the volume and the entropy of glass and liquid are indistinguishable, i.e., at δT and $\delta p \rightarrow 0$, $V_{Tg} = G_p$ and $S_{Tg} = -G_T$. Thus a sufficient condition for the glass transition is

$$\Delta G = G_{d\xi=0} - G_{A=0} \geq 0 \quad (50)$$

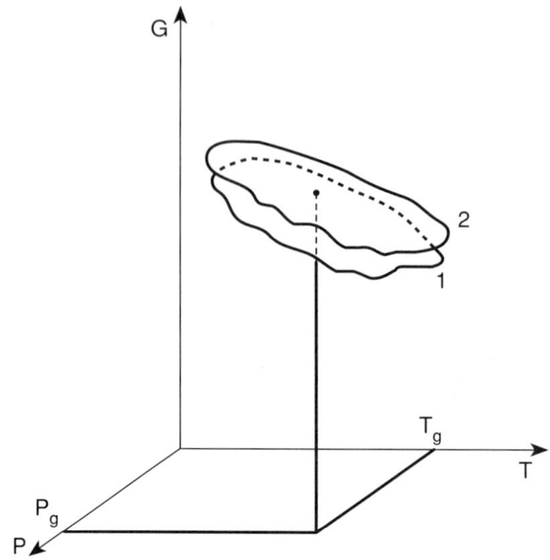

FIGURE 2. Contact of the thermodynamic potential surfaces at the glass transition. 1, liquid; 2, glass.

This means that the free energy surface corresponding to glass contacts with the free energy surface of the equilibrium liquid, at least at one point (T_g,p_g) — see Figure 2. The total set of distances between the surfaces shown in Figure 2 also defines a surface of ΔG, and $\Delta G = 0$ at the point (T_g,p_g) (a similar model is given in Reference 17).

Our task is now reduced to ascertaining the signs of the increments of the thermodynamic potential second derivatives for which glass transition is possible, i.e., the sign yielding the correct inequality in relation (50).

The above situation can be explained by means of a mechanical model. Imagine a cone suspended by a thread attached to the center of its base. We let the cone descend by increasing the length of the thread. The potential energy of the cone thus far is determined solely by the length of the thread. If there is a plane beneath the cone, its vertex at a certain moment will touch this plane. When the thread length is increased further, the position of the cone becomes unstable. If the thread is regarded analogously to temperature or pressure, the condition of the cone vertex touching the plane must correspond to the equal sign in relation (50). As soon as the affinity of the vertically hung cone to its new inclined position arises (i.e., the thread length increases with respect to the contact position, δT, $\delta p \neq 0$), the inequality sign appears in relation (50).

E. THERMODYNAMIC CRITERIA FOR THE GLASS TRANSITION

The search for the conditions of a ΔG surface minimum is equivalent to finding thermodynamic criteria for the glass transition. The expression for ΔG

derived from (50) has the following form:

$$\Delta G = G_{d\xi=0} - G_{A=0} = (-1/2)\left(G_{\xi p}\delta p\delta\xi + G_{\xi T}\delta T\delta\xi\right) \tag{51}$$

and is correct at the specified conditions of tangency, i.e., at $G_p = V$, $G_T = -S$ being single-valued at the tangency point. (In deriving this, it is presumed that for the equilibrium state $G_\xi = 0$ and, since $G_\xi = G_{\xi T}\delta T + G_{\xi p}\delta p + G_{\xi\xi}\delta\xi$, G_ξ is expressed via $G_{\xi T}$ and $G_{\xi p}$). The relation (51) appears to include only those characteristics (properties) that belong to the initial equilibrium state of liquid, i.e., $G_{\xi p}$ and $G_{\xi T}$ at T_g. From (51) we find that in the neighborhood of the point (T_g,p_g) the deviations of entropy, volume, and affinity from their equilibrium values are

$$\Delta S = -\left(\partial(\Delta G)/\partial T\right)_p = (1/2)\left[G_{\xi T}d\xi + G_{\xi T}\left(\partial\xi/\partial T\right)_p\delta T\right] \tag{52}$$

$$\Delta V = \left(\partial(\Delta G)/\partial p\right)_T = (-1/2)\left[G_{\xi p}\delta\xi + G_{\xi p}\left(\partial\xi/\partial p\right)_T\delta p\right] \tag{53}$$

and

$$\begin{aligned}\Delta G_\xi = \left(\partial(\Delta G)/\partial\xi\right) = (-1/2)\,[&G_{\xi p}\left(\partial p/\partial\xi\right)_T\delta\xi + G_{\xi p}\delta p \\ &+ G_{\xi T}\left(\partial T/\partial\xi\right)_p\delta\xi + G_{\xi T}\delta T]\end{aligned} \tag{54}$$

where the terms involving third derivatives of G are neglected.

One can now plot an auxiliary function $\delta(\Delta G)$ which is expressed in terms of the derivations of the entropy, volume, and affinity (ΔS, ΔV, and ΔG_ξ) of the nonequilibrium state (glass) from their values at the point (T_g,p_g), where ΔS, ΔV, and $\Delta G_\xi = 0$:

$$\delta(\Delta G) = -\Delta S\delta T' + \Delta V\delta p' + \Delta G_\xi\delta\xi' \tag{55}$$

This function, as one can easily see when $\delta T' = \delta T$, $\delta p' = \delta p$, and $\delta\xi' = \delta\xi$ approach zero, is proportional to ΔG [Equation (51), exactly, $\delta(\Delta G) = 4\Delta G$]. Therefore, finding the conditions of ΔG minimum in this situation is equivalent is equivalent to finding those of $\delta(\Delta G)$ minimum.

It is important to take into account that at $\delta T \to 0$ and $\delta p \to 0$ thermodynamic restrictions superimposed on the equilibrium state ($A = 0$ or $\delta(\Delta G) = 0$) and on the frozen state ($\delta\xi = 0$) must simultaneously be maintained. The reason for this is that the states of liquid and glass at the point (T_g,p_g) are indistinguishable. Then from (51)–(55) at $\Delta G = 0$, $\delta(\Delta G) = 0$, and at $\delta\xi = 0$, we have

$$-\frac{\left(\partial p/\partial\xi\right)_T}{\left(\partial T/\partial\xi\right)_p} = \frac{\delta p_g}{\delta T_g} = \frac{\Delta S}{\Delta V} = -\frac{G_{\xi T}}{G_{\xi p}} \tag{56}$$

On the other hand, the condition for the existence of the equilibrium state requires $\delta G_\xi = 0$. For the equilibrium state, at every point $G_\xi = 0$, also.

Therefore, writing δG_ξ as above,

$$\delta G_\xi = G_\xi = G_{\xi p}\delta p + G_{\xi T}\delta T + G_{\xi\xi}\delta\xi \tag{57}$$

we have

$$\delta\xi = -\frac{G_{\xi p}}{G_{\xi\xi}}\delta p - \frac{G_{\xi T}}{G_{\xi\xi}}\delta T \tag{58}$$

Obviously, only one of the three variations of the variables T, p, and ξ is independent [via the relationships of Equations (56) and (58)]. The condition $\delta\xi = 0$ that is additionally superimposed on the relation (58) at the glass transition point also requires that the relation between δp_g and δT_g have the form of (56).

The relation

$$\delta\xi = \left(\frac{\partial\xi}{\partial p}\right)_T \delta p + \left(\frac{\partial\xi}{\partial T}\right)_p \delta T \tag{59}$$

is equivalent to equation (58). By differentiating (58) from (52)–(54), and (59), at $\delta\xi \to 0$, we have

$$(\partial\xi/\partial p)_T = -G_{\xi p}/G_{\xi\xi}, \quad (\partial\xi/\partial T)_p = -G_{\xi T}/G_{\xi\xi} \tag{60}$$

and

$$\Delta G_\xi = \Delta V\frac{G_{\xi\xi}}{G_{\xi p}} - \Delta S\frac{G_{\xi\xi}}{G_{\xi T}} \tag{61}$$

The results obtained could be formulated as follows: Variation of glass transition pressure as a function of glass transition temperature variation is defined only by the properties of liquid under selected conditions (T and p), namely, by the relations (56) and (58) or (56) and (60). Of the three variables whose variations may be specified arbitrarily (T, p, and ξ), only one is independent. The source of the dependence of some variables upon others lies in thermodynamic restrictions that are simultaneously superimposed on the system state in the glass transition point, i.e., $A = 0$, ξ = constant. The thermodynamic function $\delta(\Delta G)$, the conditions for the minimum of which we are seeking (these conditions being equivalent to those of ΔG minimum), is therefore a complex function. The function $\delta(\Delta G)$ depends on the other three functions (ΔS, ΔV, and ΔG). These latter three, under glass transition conditions, are connected by two relations characterizing the liquid state [the relations (56) and (61)]; that is why only one of them is independent.

The minimum of a continuous, single-valued function requires that the determinant formed by the second derivatives of that function be positive.[23] Considering the complex function $\delta(\Delta G)$, we construct a determinant which is a functional determinant (Jacobian)

$$J = \begin{vmatrix} \frac{\partial^2(\delta(\Delta G))}{\partial T^2} & \frac{\partial^2(\delta(\Delta G))}{\partial T \partial p} & \frac{\partial^2(\delta(\Delta G))}{\partial T \partial \xi} \\ \frac{\partial^2(\delta(\Delta G))}{\partial p \partial T} & \frac{\partial^2(\delta(\Delta G))}{\partial p^2} & \frac{\partial^2(\delta(\Delta G))}{\partial p \partial \xi} \\ \frac{\partial^2(\delta(\Delta G))}{\partial \xi \partial T} & \frac{\partial^2(\delta(\Delta G))}{\partial \xi \partial p} & \frac{\partial^2(\delta(\Delta G))}{\partial \xi^2} \end{vmatrix} > 0 \tag{62}$$

When expressing the second derivatives through the relations (52)–(55), we have

$$J = \begin{vmatrix} -G_{\xi T}\left(\frac{\partial \xi}{\partial T}\right) & -G_{\xi T}\left(\frac{\partial \xi}{\partial p}\right) & -G_{\xi T} \\ -G_{\xi p}\left(\frac{\partial \xi}{\partial T}\right) & -G_{\xi p}\left(\frac{\partial \xi}{\partial p}\right) & -G_{\xi p} \\ -G_{\xi T} & -G_{\xi p} & +G_{\xi\xi} \end{vmatrix} > 0 \tag{63}$$

From the theorems of the Jacobian value, for m equations connecting n functions upon which the extremum complex function depends, we know that the values of the determinants of $n - m + 1$, $n - m + 2$ and higher orders are zero.[24] The only nonzero determinants are those of order less than $n - m$, i.e., the number of independent functions. In our case $n = 3$, $m = 2$, and therefore only diagonal elements of determinants (62) and (63) are positive simultaneously (each one forms a determinant of the first rank while the choice of the variable sequence is arbitrary):

$$-G_{\xi T}\left(\frac{\partial \xi}{\partial T}\right) > 0 \qquad -G_{\xi p}\left(\frac{\partial \xi}{\partial p}\right) > 0 \qquad G_{\xi\xi} > 0 \tag{64}$$

One can come to the same conclusion on the basis of constancy of the first two rows and of different columns in (63).

The physical meaning of the inequalities in (64) becomes clear from consideration of (47), (65), and (66):

$$\left(\frac{\partial^2 G}{\partial T^2}\right)_{p,d\xi=0} - \left(\frac{\partial^2 G}{\partial T^2}\right)_{p,A=0} = \left(\frac{\partial^2 (\delta(\Delta G))}{\partial T^2}\right)_{p,A=0}$$

$$= -G_{\xi T}\left(\frac{\partial \xi}{\partial T}\right)_{p,A=0} = \frac{\Delta C_p}{T} \tag{65}$$

and

$$\left(\frac{\partial^2 G}{\partial T^2}\right)_{T,d\xi=0} - \left(\frac{\partial^2 G}{\partial T^2}\right)_{T,A=0} = \left(\frac{\partial^2 (\delta(\Delta G))}{\partial p^2}\right)_{T,A=0}$$

$$= -G_{\xi p}\left(\frac{\partial \xi}{\partial p}\right)_{T,A=0} = V\Delta\kappa_T \tag{66}$$

where ΔC_p and $\Delta\kappa_T$ are increments of the heat capacity and the isothermal compressibility coefficient at the transition from glass to liquid; τ in (47) is the relaxation time. Thus at the glass transition caused by lowering the temperature, the heat capacity and compressibility always go down while the glass state is a thermodynamical nonequilibrium. This is possible only when the relaxation time is simultaneously nonzero (when $G_{\xi\xi}$ does not go to infinity, as it has been specified previously) and not infinite ($G_{\xi\xi} > 0$), i.e., at finite relaxation times:

$$\Delta C_p > 0 \qquad \Delta\kappa_T > 0 \qquad 0 < \tau < \infty \tag{67}$$

The sign of the thermal expansion coefficient increment

$$\Delta\alpha = \alpha(1) - \alpha(g) = G_{\xi T}\left(\frac{\partial \xi}{\partial p}\right)_T V^{-1} \tag{68}$$

at the glass transition may be either positive or negative. (Here l denotes liquid and g denotes glass).

Similar analysis for other kinds of thermodynamic potentials yields an analogous result: $\Delta C_V > 0$, $\Delta\kappa_S > 0$ and $0 < \tau < \infty$.

The inequalities of (67) are the thermodynamic criteria of glass transition that we sought. Breaking the relaxing condition for τ makes either the existence of glass or of equilibrium liquid devoid of physical meaning. That is, when $\tau = 0$ the equilibrium liquid always remains the same; when $T = \infty$ the glass would never relax up to the equilibrium liquid state (even learning of its existence becomes impossible).

The theory evidently does not impose any limitations which would result in the fact that glass as an unstable state must be obtained only from metastable liquid. However, this is exactly the situation when relaxation times are great enough to be compared with the time scales of experiment. Sharp viscosity increases for supercooled liquids provide the most favorable kinetic conditions for a glass transition.

The thermodynamic criteria of the origin of the unstable state differ in principle from the Gibbs criteria (29) by the fact that they determine the sign of the heat capacity and compressibility increments but not the sign of the very coefficients. The first of the (67) conditions is well known as the Van't Hoff empirical rule.[25]

II Prigogine-Defay Ratio and the Problem of Structural Parameter Plurality

A. GLASS TRANSITION AS COMPARED TO PHASE TRANSITIONS OF THE FIRST AND SECOND ORDER

The sign of ΔC_p at the glass transition is defined uniquely. It does not depend on the vitreous state resulting from either nonequilibrium cooling or step "super-heating" of the initial equilibrium liquid of high viscosity. A similar conclusion applies to the increase in compressibility. There is no such situation at phase transitions of the first order. In such cases the succession of changing phases is defined solely by the line of equal thermodynamic potentials (Figure 1).

The dependence of glass transition temperature vs. pressure, unlike the similar one for "ideal" phase transitions of the second order according to Ehrenfest,[26] is not associated with the existence of a line on which the free energies of phases are equal. It also follows from Relation (56) that dp_g/dT_g depends only on the $G_{\xi T}/G_{\xi p}$ ratio, i.e., on the properties of liquid. The latter differ in various points selected arbitrarily from the multitude of (T_g, p_g). Accordingly, the choice of the $\Delta G = 0$ condition is also arbitrary and defined only by kinetic conditions of freezing.

For second order phase transitions, such as "order–disorder" transitions, at $T < T_{ph.tr.}$, $d\zeta/dT \neq 0$ and at $T > T_{ph.tr.}$, $d\zeta/dT = 0$, where ζ is an ordering parameter.[27] For the glass transition, when $T < T_g$, $d\xi/dT = 0$ and when $T > T_g$, $d\xi/dT \neq 0$. At $T \to T_{ph.tr.}$ $\Delta C_p \to \infty$, and at T_g, ΔC_p is finite.

The dependence of glass transition pressure variation on the glass transition temperature variation (56) resembles the Clausius-Clapeyron equation for phase transitions of the first order. In this case ΔS and ΔV are increments of the entropy and volume at a reversible transition from one type of equilibrium state to another. In Equation (56) those values characterize an irreversible and nonequilibrium transition process from one structural state of a system to another. The values of these characteristics, as is mentioned above, depend on the properties of liquid at arbitrarily selected combinations of T_g and p_g.

Equation (56) relates the baric dependence of the glass transition temperature and the temperature dependence of the glass transition pressure with the kinetic coefficient that characterizes a substance, but subject to very rigid limitations. These are the condition $G_{\xi\xi p} = G_{\xi\xi T} = 0$ and the condition referring these dependences to one point (T_g, p_g) when $G_{\xi\xi}$, γ, and τ [Relation (47)] are uniquely related to each other:

$$\left(\partial p_g / \partial T_g\right)\Big|_\tau = \left(\partial\tau / T_g\right)_p / \left(\partial\tau / p_g\right)_T = \left(\partial\gamma / \partial T_g\right)_p / \left(\partial\gamma / \partial p_g\right)_T \tag{69}$$

These are the conditions for isokinetic examination of the dependence dp_g/dT_g where it has a simple physical meaning. When the coefficient dp_g/dT_g is investigated at other conditions, the derivative dp_g/dT_g variation may also be caused by changing the kinetic coefficient associated with the latter's dependence on temperature and pressure. Thus the phenomena of glass transitions and phase transitions may exhibit elements of formal resemblence but not of physical essence.

B. THERMODYNAMIC THEORY

The determinant (63) would be positive if all the variables (T, p, and ξ) were mutually independent. We have ascertained, however, that only one of the three variables can be chosen independently. That results in a determinant of even the second order being zero:

$$\begin{vmatrix} -G_{\xi T}\left(\dfrac{\partial\xi}{\partial T}\right) & -G_{\xi T}\left(\dfrac{\partial\xi}{\partial p}\right) \\ -G_{\xi p}\left(\dfrac{\partial\xi}{\partial T}\right) & -G_{\xi p}\left(\dfrac{\partial\xi}{\partial p}\right) \end{vmatrix} = 0 \tag{70}$$

This condition corresponds to the mutual equality of the cross products of the determinant (70) elements. Taking into account the meaning of these elements from Equations (65), (66), and (68), and also the fact that $G_{\xi p}(\partial\xi/\partial T)_p = G_{\xi T}(\partial\xi/\partial p)_T$, we have (at $T = T_g$ and $V = V_{Tg}$)

$$\Pi \equiv \frac{\Delta C_p \, \Delta\kappa_T}{T_g V_{T_g} (\Delta\alpha)^2} = 1 \tag{71}$$

This relation is known as the Prigogine-Defay Ratio.[10] It establishes the association between the jumps of the second derivatives of a thermodynamic potential at the glass transition. It looks like the well-known Ehrenfest relation for the jumps of those very coefficients for ideal phase transitions of the second order. However, as we saw in the previous section, the nature of the internal changes of the system at the glass transition differs from that at phase transitions.

New possibilities are revealed for the case when the internal structure of the phase undergoing glass transition is defined by not only one but by several mutually independent parameters. Independent structure parameters are those that contribute independently to the entropy and volume:

$$\begin{aligned}&(\partial V/\partial \xi_1)/(\partial S/\partial \xi_1) \neq (\partial V/\partial \xi_2)/(\partial S/\partial \xi_2) \neq \cdots \\ &\neq (\partial V/\xi_k)/(\partial S/\partial \xi_k)\end{aligned} \tag{72}$$

This means that among all the parameters, k in number, there are no proportional relations. When several parameters are available (ξ_k), the rank of the determinant (63) increases and its first two rows are written as

$$C_p/T - V\Delta\alpha - G_{\xi_1 T} - G_{\xi_2 T} - \cdots - G_{\xi_i T} - \cdots - G_{\xi_k T} \tag{73}$$

and

$$-V\Delta\alpha + V\Delta\kappa_T - G_{\xi_1 p} - G_{\xi_2 p} - \cdots - G_{\xi_i p} - \cdots - G_{\xi_k p} \tag{74}$$

where

$$\Delta C_p/T = \sum_{i=1}^{i=k} G^2_{\xi_i T} \cdot G^{-1}_{\xi_i \xi_i} \tag{75}$$

$$\Delta\kappa_T = V^{-1} \sum_{i=1}^{i=k} G^2_{\xi_i p} \cdot G^{-1}_{\xi_i \xi_i} \tag{76}$$

and

$$\Delta\alpha = -V^{-1} \sum_{i=1}^{i=k} G_{\xi_i T} \cdot G_{\xi_i p} \cdot G^{-1}_{\xi_i \xi_i} \tag{77}$$

Accordingly, increases in the rank of the determinant may be positive.

The expression for Π has the following form:

$$\Pi = \sum_{i=1}^{i=k} \left(G_{\xi_i T} G^{-1}_{\xi_i \xi_i}\right) \sum_{i-1}^{i=k} \left(G^2_{\xi_i p} G^{-1}_{\xi_i \xi_i}\right) \left[\sum_{i=1}^{i=k} \left(G_{\xi_i T} G_{\xi_i p} G^{-1}_{\xi_i \xi_i}\right)^2\right]^{-1} \tag{78}$$

Since all $G_{\xi_i \xi_i} \neq 0$ (all the parameters can relax), we apply the well-known Cauchi-Schwarz inequality:[28] the product of the sums of the values squared is more than the sum of the squares of their products if the values are not proportional to each other. Here

$$G_{\xi_1 T} / G_{\xi_1 p} \neq G_{\xi_2 T} / G_{\xi_2 p} \neq \cdots \neq G_{\xi_i T} / G_{\xi_i p} \neq \cdots \neq G_{\xi_k T} / G_{\xi_k p} \tag{79}$$

[see Equation (72)]. Therefore, $\Pi > 1$ if the indicated inequalities are valid for any $k > 1$. Thus, if the number of independent parameters defining the liquid structure is more than 1, the jumps of thermodynamic coefficients at T_g are certainly bound by the inequality

$$\Pi \equiv \frac{\Delta C_p \Delta \kappa_T}{T_g V_{T_g} (\Delta\alpha)^2} > 1 \tag{80}$$

This relation was pointed out for the first time by Meixner[29] in his discussion with Davies in 1952. Later that relation was examined in many studies. Similar, but not always analogous, relations may be obtained for other potentials differing in the way the thermodynamic parameters (V,S), (p,S), and (V,T) are defined.[14]

In fairness, it should be noted that DiMarzio[30] denied that the concept of independent structure parameters was applicable to the description of the glass transition. But he postulated initially that the line dT_g/dp_g is defined uniquely. Mathematically that postulate cannot yield any other result than $\Pi = 1$, to which the only structure parameter corresponds also to Ehrenfest's theory of ideal phase transitions of the second order.[26]

The most striking examples of structure parameters depending slightly on each other are those defining dipole reorientation and those defining structural changes due to temperature variations. If the thermodynamic variables are electric field E and temperature T, for system stability, as Moynihan and Lesikar[31] have shown, the relation

$$\Pi \equiv \frac{V \Delta C_p \Delta\varepsilon}{\varepsilon_0 E^2 \left\{\Delta\left[\partial(\chi V)/\partial T\right]\right\}^2} \geq 1 \tag{81}$$

must hold, where χ is the electric susceptibility, ($\chi = \varepsilon - 1$), $\Delta\varepsilon$ is the relaxational part of the dielectric constant (the difference between ε values for the lowest and for the highest frequencies), ε_0 is the dielectric permeability of free space, and V is the volume. At $E \to 0$, $\Pi \to \infty$. This means that in the absence of a field an isotropic liquid cannot be polarized by a temperature jump. $\Pi \to 1$ only at the E values exceeding the dielectric strength of the material. A similar result was obtained by the same authors[31] for the electromechanical stability condition (E and p variable). In both cases for nonzero E values, $\Pi > 1$, which obviously points to the possibility of partial depolarization by structural changes produced by the temperature and pressure variations.

If a system is inhomogeneous, then, as Gupta and Haus[32] and Shultz and Filippov[33] have shown, Π is always > 1, since the factor of variable concentration defining phase equilibrium behaves as an additional parameter.

The above relations are of extreme importance for investigating the nature of the glass transition. Thermodynamics gives a convincing criterion for the experimental verification and conformity of the human-created speculative models of liquid structure with reality.

C. EXPERIMENT

To calculate Π, one would need knowledge of the temperature dependence of heat capacity, compressibility, density (and volume thermal expansion coefficient) both for glass below the glass transition temperature and for metastable liquid above the glass transition temperature. These properties are usually studied in various laboratories differing by their specialization fields. This circumstance complicates the use of experimental data from various sources, first because of possible differences in chemical purity of the substances examined. Dynamic regimes of the temperature variations also may be different. Therefore all results require calculations of the coefficient increments for a uniform (for every substance) glass transition temperature. For adequate comparison of different substances, it is evidently best to take for the glass transition temperature the T_g that corresponds to the viscosity of $1 \cdot 10^{13}$ P. The most reliable information about heat capacity is obtained by direct calorimetric experiment, where the heat capacity upsurge a little above T_g is excluded. It has relaxational nature and is characteristic only of the data obtained in the regime of increasing sample temperature.* Data on thermal expansion are treated similarly if they were obtained in the heating regime. The jump in the compressibility coefficient is usually calculated from measurements of ultrasound velocities at various frequencies (beyond the dispersion region) by means of extrapolating the data for solid glass and for its melt. Maximum errors for ΔC_p, $\Delta\kappa_T$, and $\Delta\alpha$ evaluations are no more than 5, 10, and 15%, respectively. A detailed description of the procedure for these calculations, based on different sources, is given in Reference 34.

Davies and Jones,[14] using the most reliable available data, calculated the values of Π for Se (2.4, $T_g = 300$ K), glucose (3.7, $T_g = 300$ K), rubber (8.3, $T_g = 320$ K), and polystyrene (16, $T_g = 350$ K). Gupta and Moynihan[17] obtained $\Pi = 4.7$ ($T_g = 550$ K) for B_2O_3, $\Pi = 4.5$ ($T_g = 340$ K) for a nitrate mixture of $0.4Ca(NO_3)_2 \cdot 0.6KNO_3$, and $\Pi = 2.2$ ($T_g = 303.8$ K) for polyvinylacetate. Nemilov et al.[34] have calculated Π for silicate and borate glasses. For several silicate glasses of complex composition (a windowpane glass, a soda-lime-silicate glass, N 710 NBS USA, and several crown, flint, and heavy barium

* The nature of this phenomenon is discussed later.

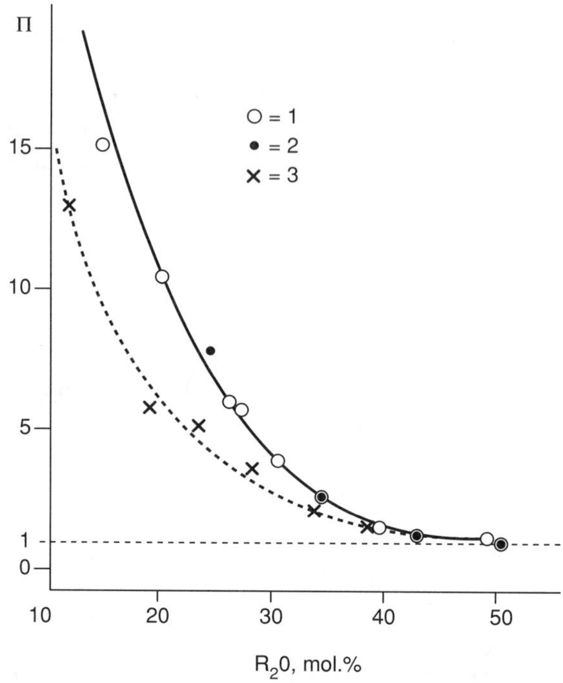

FIGURE 3. The value of the Prigogine-Defay Ratio for $Na_2O–SiO_2$ (1,2) and $K_2O–SiO_2$ (3) systems. 1, 2 correspond to different literature sources on the compressibility. (From Nemilov, S. V., Bogdanov, V. N., Nikonov, A. M., Smerdin, S. N., Nedbai, A. I., and Borisov, B. F., *Fiz. i Khim. Stekla (Sov. J. Glass Phys. Chem.),* 13, 801, 1987. With permission.)

crown optical glasses), the Π values were within the limits of 2 to 2.7 at T_g varying from 727 to 921 K. Similar values for Π are reported by Moynihan and Lesikar[31] for two other glasses of similar type.

Of utmost interest is the change of Π values for systematic variations in the composition of silicate and borate glass systems.[34] These systems are considered as models, and information on their structure is usually the most reliable. Figure 3 shows results for changes in the $Na_2O–SiO_2$ and $K_2O–SiO_2$ systems. For pure SiO_2 glass, $\Pi \sim 1 \times 10^3$ (Reference 34) or even 2×10^5 (Reference 35). The main cause of the high Π values for this substance is small variations in the values for the volume expansion coefficient at the glass transition whereas changes in the heat capacity and compressibility are not very different from those of other glasses. Evidently, at variations in melt composition and structure, the Π-value decreases gradually and goes to 1 for metasilicate compositions ($R_2O \cdot SiO_2$, 50 mol.% of R_2O). The latter melts readily crystallize. A similar tendency of decreasing Π with the growth of oxide (Na_2O or K_2O) content over 20 mol.% is observed in $R_2O–B_2O_3$ systems, but because of crystallization, compositions containing more than 37 mol.% R_2O were not studied (Reference 34).

Thus the available experimental data prove that at the glass transition not only one but two or more mutually independent structural parameters are being frozen. Volume relaxation is expected to proceed irrespective of the relaxation of heat capacity (entropy).

D. SIMPLEST MODELS OF STRUCTURAL PARAMETERS

The problem of creating a model of glassforming liquid structure that will yield values of $\Pi > 1$ is one of the most difficult problems in vitreous state theory and has, of course, heuristical sense.

The simplest model is the theory of free volume developed for the glass transition by Hirai and Eyring.[36] Phonons can interact with each other, giving rise to holes; the holes can annihilate giving rise to phonons. In the solid state below T_g the concentration of holes is assumed to be frozen and their diffusion to be slow. If ε_h is the energy of a hole formation, v_h is its volume, and v_0 is the volume of a molecule (or of a group of atoms), then the excess heat capacity, volume expansion, and compressibility of liquid in comparison with those of glass are expressed by following relations:

$$\Delta C_p = \left(v_0 / v_h\right)\left(\varepsilon_h / k_B T\right)^2 R \exp\left[-\left(\varepsilon_h + p v_h\right) / k_B T\right] \tag{82}$$

$$\Delta\alpha = \left(\varepsilon_h / k_B T^2\right) \exp\left[-\left(\varepsilon_h + p v_h\right) / k_B T\right] \tag{83}$$

$$\Delta\kappa_T = \left(k_B T / v_h\right)^{-1} \exp\left[-\left(\varepsilon_h + p v_h\right) / k_B T\right] \tag{84}$$

The theory gives the result $\Pi = 1$. This is understandable because the variations of both volume and enthalpy are uniformly correlated with the concentration of holes and, therefore, are associated to each other.

Kauzmann[37] suggested that at T_g certain molecular degrees of freedom are frozen in. They are neither rotational nor translational. The latter can't contribute significantly to the system's potential energy. Such degrees of freedom may be only configurational ones defining differences between atomic (or molecular) arrangements. It should be mentioned that even in the 1940s such ideas were suggested by many authors. For example, in the study by Douglas[38] it was supposed that in silicate glasses, even in their solid state, each bridge oxygen occupies one of the two positions, (1) or (2), differing by inversion transfer over a small potential barrier

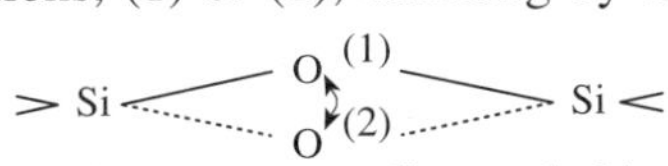

In the liquid state both positions are equally probable. Obviously, the temperature at which different positions of oxygen can be frozen in depends on

the magnitude of the potential barrier. Such two-level models are applied to interpret the results of nuclear spin-lattice relaxation in inorganic glasses below T_g. For example, for B_2O_3 an energy barrier value of 4.4 kJ per mole was obtained (Rubinstein et al.[39]).

Gibbs and DiMarzio,[40,41] Gibbs,[42] Vol'kenshtein[43] evolved configurational statistics for polymer chains. The internal rotation of structural groups (such as $-CH_2-$ in hydrocarbons) in the liquid state results in higher values of heat capacity, thermal expansion, etc.

Such ideas are attractive starting points for understanding the nature of several independent structural parameters. The simplest one is the model of two states differing by their structure and by dynamic equilibrium. That model is equivalent to only ξ and leads inevitably to a value of $\Pi = 1$ (Reference 10). Notwithstanding the different physical meanings of those states implied by various authors (e.g., References 44–47), the expression for the excess configuration heat capacity of an equilibrium liquid (with respect to a liquid of unchanging structure) has the form of a well-known function by Schottky[10]

$$\Delta C_p / R = (\Delta G / RT)^2 \exp(\Delta G / RT)\left[1 + \exp(\Delta G / RT)\right]^{-2} \tag{85}$$

where ΔG is the free energy change at the transformation and $R = Nk_B$ is the gas constant. The suggestion of the existence of isomerization reactions has been made by Davies and Lamb[44] and in later works by various authors.

We have tested the applicability of the Schottky function to a number of inorganic and molecular systems.[20] The function (85) does not render the shapes of experimental ΔC_p curves over a wide temperature range correctly, and the difference between the absolute values may be 10 times as large. The model of several bound excited states (e.g., dangling bonds) complicates the above expression but brings it nearer to the experiment (constant ΔG_i and ΔS_i are selected) (Angell and Rao[46]). It is quite obvious that to explain $\Pi > 1$, it is necessary to introduce a more complex model and a method for modeling both heat capacity and volume changes such that the direct connectedness of states disappears.

E. GENERAL MODELS OF TWO INDEPENDENT STRUCTURAL PARAMETERS

Leidecker et al.,[48] based on the results of heat capacity, compressibility, and thermal expansion studies of B_2O_3, proposed that in the liquid state there exists a doubly degenerated excited state. Thus for the same energy excess ΔU (with respect to the ground state), states are possible that differ only by their volume (ΔV). All states are bound by dynamic equilibrium. At atmosphere pressure ($p\Delta V \ll \Delta U$) the coefficient increments are described by equations

$$\Delta C_p = \left[(\Delta U)^2 / RT^2\right] x_1 (1 - x_1) \tag{86}$$

$$\Delta\alpha = \left(\delta V \cdot \Delta U / RT^2\right) x_1 (1 - x_1) \tag{87}$$

$$\Delta\kappa_T = \left[(\delta V)^2 / RT\right](r + x_1)(1 - x_1) \tag{88}$$

where $\delta V = \Delta V(1 + r)$ is the effective volume change, r is the relative populations of the levels in the double-degenerated state ($r = x_2/x_3$), and

$$x_1 + x_2 + x_3 = 1 \tag{89}$$

$$x_1 = \left\{1 + \exp\left[(\Delta U - T\Delta S)/RT\right]\right\}^{-1} \tag{90}$$

$$x_2 = r(1 - x_1)/(1 + r) \tag{91}$$

$$x_3 = (1 - x_1)/(1 + r) \tag{92}$$

$$r = \exp(p\Delta V / RT) \tag{93}$$

Experimental data for the temperature dependence of the thermodynamic coefficients, including the relaxational part of the compressibility above 700 K, are approximated by the above relations at selected values of $\Delta U = 21.3$ $kJ \cdot mol^{-1}$, $\Delta S = 28.3\ J \cdot mol^{-1}K^{-1}$, $\delta V = 0.117\ cm^3g^{-1}$, and $r = 3.8$. In this model $\Pi = 1 + r/x_1 > 1$ and at 700 K $\Pi = 8$; with increasing temperature Π increases. The authors suggested the ground state structure corresponding to a tightly packed system of directed B–O bonds. An excited state may be produced in two ways: bonds rupture and the configuration is conserved, or bonds rupture and the structure loosens. At T_g the equilibrium between states is frozen, and contributions into the heat capacity and other coefficients disappear. Here the independent structure parameters are the excited state concentrations and the relative populations of the levels of doubly degenerated states r. It is the latter that defines the disproportion of the entropy and volume changes.

Roe[49] suggested another model to explain the appearance of $\Pi > 1$. He considered a one-dimensional quasi-lattice structure consisting of both empty and occupied "molecule" sites. The number of empty sites is n. Occupied sites, of total number N, form clusters. They are characterized by the number m, the number of uninterrupted sequences of occupied sites. The volume V of the system is defined by the relation

$$V = v(N + n) \tag{94}$$

where v is the volume of each site, N + n is the total number of occupied and empty sites, and v is assumed to be the same for occupied and empty sites. The configurational energy of the system is defined by the relation

$$E = 2m\varepsilon \tag{95}$$

where ε is the energy contributed by each pair of neighboring occupied and unoccupied sites, m is the number of sites in an uninterrupted sequence of occupied sites. The numbers m and n behave as two independent structure parameters defining ξ_1 and ξ_2.

For an isothermal-isobaric ensemble of such systems having large N values, the partition function may be expressed as

$$Z(T,p,m,n) = [j(T)]^N \Omega(m,n)\exp(-E/k_B T)\exp(-pV/k_B T) \tag{96}$$

where j (T) is the internal partition function of individual "molecules". For the chain structure described, Ω (m,n) is the degree of degeneracy of the states specified by m and n:

$$\Omega(m,n) = \frac{N!\,n!}{(N-m)!\,(n-m)!\,(m!)^2} \tag{97}$$

Since the Gibbs free energy may be defined as

$$G = -k_B T \ln[Z(T,p,m,n)] \tag{98}$$

we have

$$G = -Nk_B T \ln j(T) + 2m\varepsilon + v(N+m)p - k_B T \ln \Omega(m,n) \tag{99}$$

For large N, m, and n values, application of the Stirling formula leads to

$$\ln \Omega = N \ln N + n \ln n - (N-m)\ln(N-m)$$
$$-(n-m)\ln(n-m) - 2m \ln m \tag{100}$$

For an equilibrium system, m and n values are found from the G extremum condition. Using Relations (99) and (100) $-G_m$ and $-G_n$ can be found immediately. These are equivalent to the values of affinity for differing parameters of the system order:

$$-G_m = -\frac{\partial G}{\partial m} = -2\varepsilon + k_B T \cdot X \tag{101}$$

$$X = \frac{\partial \ln \Omega}{\partial m} = \ln\left[\frac{(N-m)(n-m)}{m^2}\right] \tag{102}$$

$$-G_n = -\frac{\partial G}{\partial n} = vp + k_B T \cdot Y \tag{103}$$

$$Y = \frac{\partial \ln \Omega}{\partial n} = \ln\left(\frac{n}{n - m}\right) \tag{104}$$

From the $G_m = 0$ and $G_n = 0$ conditions for the case of an equilibrium liquid, it is found that

$$n/(n - m) = \exp\left(pV/k_B T\right) \tag{105}$$

$$(N - m)(n - m)/m^2 = \exp\left(2\varepsilon/k_B T\right) \tag{106}$$

These equations, together with (94) and (95), define numerical relations connecting n, m, and N.

The entropy S and volume V values for the glass transition point, based on considerations from the previous sections, are obtained as:

$$S = -\left(\frac{\partial G}{\partial T}\right)_{p,m,n} = N\frac{\partial}{\partial T}\left(k_B T \ln j\right) + k_B \ln \Omega \tag{107}$$

$$V = \left(\frac{\partial G}{\partial p}\right)_{T,m,n} = v/(N + n) \tag{108}$$

There is no sense in continuing this analysis further. Relations (107) and (108) show clearly that there must be no proportionality in the relations of the volume and entropy derivatives for various structure parameters equivalent to n and m, and that Relation (80) must work. This could have been anticipated because from the very beginning the functional dependencies of volume and entropy [see Relations (94) and (95)] on these parameters were different.

Thus, Roe's model[49] also explains the occurrence of more than one structural parameter and a value of Π>1. In the opinion of the author of this theory, the manifestation of the two independent parameters can, from the kinetic point of view, be interpreted as follows. When liquid is gradually cooled initially only one parameter can be frozen in. For instance, the free volume of the system can reach a certain critical value for the glass transition fraction of the total volume. At this moment the ability of redistribution of the free volume inside the system may persist. At lower temperatures this process must be frozen too. In this case, in principle, the value of $\Pi = 1$ might be observed for every T_g. However, that would require division of the ΔC_p jump (as well as $\Delta\kappa_T$ and $\Delta\alpha$) into two parts, according to the different dependencies of the second derivatives of G on m and n. So far, however, such separation could not be successfully demonstrated in experiments. The calculation of Π values from the experiment operationally corresponds to simultaneous freezing of all the parameters.

The models described above are, however, too abstract to establish for certain an association of their parameters with a definite structure of the glassforming melt on a molecular level.

F. BASIC MODEL CONCEPTS OF GLASS STRUCTURE*

In order to discuss the possible cause of the changes of Π values with composition variations in silicate glasses, it is necessary to review a certain minimum information about the structure of these glasses.

The tendency of various substances to form glass was explained by Zachariasen[2] in terms of the possibility of specific linking of elementary structural polyhedra. Thus in vitreous silica the elementary structural polyhedron is a SiO_4 tetrahedron with a silicon atom in its center. These tetrahedra are linked by their vertices by means of Si–O–Si bonds resulting in the formation of a three-dimensional spatial network. In vitreous boric anhydride, BO_3 triangles with a boron atom in their centers are linked with neighboring ones by three bridge oxygen atoms, thereby forming an almost planar structure of continuous B–O–B bonds. In this structure approximately two thirds of the boron atoms form the so-called boroxol groups (like rings):

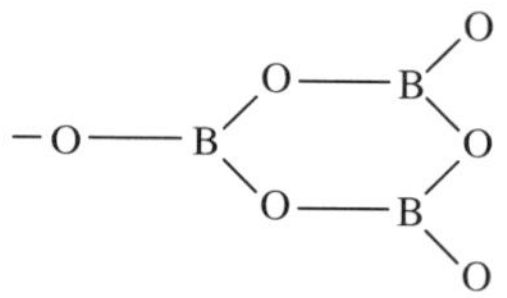

The rest of the boron atoms involved in $BO_{3/2}$ groups are distributed freely (see Mozzi and Warren[50] and Prabakar et al.[51]). The B–O bond distance is 1.37 Å. The distance between the layers is ~4.3 Å (Reference 52). The quantitative ratio of boroxol rings to $BO_{3/2}$ groups is different in the works of various authors.

Linkage of polyhedra by their edges or corners in other substances (ZrO_2, Al_2O_3, etc.), known from crystallochemistry, promotes glass crystallization. Myuller,[53] Smekal,[54] Douglas,[38] Syrkin and Dyatkina[55] and other authors noted long ago the special role of covalent (partly ionic) bridge bonds of Si–O–Si, B–O–B, etc. in glass formation. The short-range effectiveness of covalent (partly ionic) forces in bridging bonds preserve the Si–O (1.61 Å) and B–O (1.37 Å) distances and promotes a slight distortion of the Si–O–Si, B–O–B, etc. angles. Because of this, a disordered liquid structure is easily conserved when supercooled, crystallization being inhibited by the fact that the regular orientation of polyhedra requires correlated rotations of structural units. These

* The subject of this subsection is necessary not only to comprehend model interpretations of the structural parameter plurality, but also to understand many of the following sections. This information does not constitute a comprehensive analysis, however.

rotations are possible only when bridging bonds are broken or switched over while the bridging atoms change their partners — atoms of silicon, boron, etc. According to recent X-ray diffraction investigations of vitreous silica structure at increased pressure up to 10 GPa (Meade et al.[56]), an elementary tetrahedron SiO_4 unit does not change. Its volume remains constant due to high stability of O–Si–O angles and Si–O distances within a tetrahedron. Only the angles of bridging bonds undergo essential changes. These angles define the tightness of the structural package on the whole.

Introducing alkali oxides into silica reduces the number of bridging bonds that a SiO_4 polyhedron can form with its neighbors. If a bridging oxygen atom is designated by the $O_{1/2}$ symbol (it belongs to two neighboring tetrahedra), the reaction of alkali–silica interaction is expressed by equations of the form

$$2\,SiO_{4/2} + Na_2O \rightarrow 2\,Na^{+}O^{-}SiO_{3/2} \quad \text{(disilicate)} \tag{109}$$

and

$$SiO_{4/2} + Na_2O \rightarrow Na_2^{2+}O_2^{2-}SiO_{2/2} \quad \text{(metasilicate)} \tag{110}$$

Here O^- is a nonbridging oxygen atom connected with a silicon, its charge being compensated by a Na^+ cation charge. Thus, by progressive introduction of an alkali into silica, tetrahedra ionization occurs and the number of bridging bonds per each tetrahedron decreases. Layer structures are formed when every tetrahedron is linked with its neighbors by three bridging oxygen atoms (a type of disilicate); chain structures are formed when a tetrahedron is connected with its neighbors only by two bridging atoms (a type of metasilicate) and, eventually, island-like and completely ionized structures of $[SiO_4]^{4-}$ tetrahedra (a type of orthosilicate). The pointed structure type takes place spatially due to molecular polymerization. For example, in SiO_2 six-membered rings consisting of tetrahedra, not lying in one plane, may be conditionally distinguished. In silicates more complicated groupings may exist, such as $(Si_4O_{10})^{4-}$, $(Si_2O_6)^{4-}$, $(Si_3O_9)^{6-}$, etc. This problem is not discussed here. We only note that structural investigations of silicate glasses using X-ray diffraction confirm the schemes described above (e.g., References 57 and 58).

Reduction of the spatial connectivity of the framework and, especially, increases of bond ionicity lead, in this framework, to decreases of viscosity and glass transition temperatures. This also results in an increasing tendency for crystallization.

In the case of borate glasses, the addition of an alkali (or alkali-earth) oxide to boric anhydride results in partial conversion of boron atoms from threefold coordination to the fourfold coordination, e.g.,

$$BO_{3/2} + 1/2\,Na_2O \rightarrow \left[BO_{4/2}\right]^{-} + Na^{+} \tag{111}$$

The negative charge of the $[BO_{4/2}]^-$ complex is compensated by the charge of a nearby localized cation. The B–O distance in a BO_4 tetrahedron is 1.48 Å. Combining $BO_{3/2}$ and $BO_{4/2}$ groups results in the formation of complex groups, such as pentaborate $[B_5O_8]^-$, triborate $[B_3O_5]^-$, diborate $[B_4O_7]^{2-}$, etc. The growth of spatial connectivity is accompanied by increasing T_g. At alkali concentrations of more than ~30 mol.%, the number of $BO_{4/2}$ groups decreases (Griscom[59]). For equal numbers of B_2O_3 and alkali oxides, metaborates are formed with chain or ring structures of $[BO^- O_{2/2}]_n$ groups. Those compositions crystallize.

Similarly, when alumina is introduced into alkali-silicate glasses, $[AlO_{4/2}]^-$ groups are formed. In alkali-germanate glasses, groups of six-coordinated germanium are also formed; i.e., a part of germanium atoms increases their coordination number from 4 (as it is in vitreous GeO_2) to 6.

From investigations of thermodynamic properties of melts, the thermodynamic stability of various groups can be calculated and compared their quantities. It appears, for example, that in the $Na_2O–SiO_2$ system at 1273 K the most stable structures with respect to dissociation are the metasilicates and orthosilicates, and the most dissociated (about 30%) is disilicate.[60] In the $K_2O–SiO_2$ system the most dissociated is tetrasilicate $K_2O \cdot 4SiO_2$ (about 55%), disilicates are more stable (dissociated for about 20%), and the most stable are metasilicates and orthosilicates.[60] Dissociation of compounds results in the formation of the most stable compounds of nearby compositions. For example, sodium disilicate dissociation yields silica and metasilicate.

From NMR measurements of ^{29}Si nuclei, the fraction of tetrahedra that have one, two, and four bridging oxygen atoms can be separately determined for every glass composition.[61,62] These results are not only well reproduced but are also in complete agreement with thermodynamic calculations.[60]

Figure 4 shows the dependence of the *average* number of nonbridging oxygen atoms per silicon tetrahedron vs. composition. The line corresponds to the experimental error and divergence of the results for Li_2O–, Na_2O–, and $K_2O–SiO_2$ systems. It is quite obvious that the average connectivity of a silicate framework may be calculated as a single-valued function of composition.

Combined investigation of frequency and temperature dependencies of ultrasound velocity and attenuation and shear viscosity in the same systems (1–6 MHz, 1070–1570 K) (i.e., Nikonov et al.[63]) are also of interest in connection with the above structural ideas. In the composition region of 12–40 mol.% R_2O (Na_2O or K_2O), increasing the alkali oxide concentration narrows the time relaxation spectrum and the most probable and the average relaxation times, and reduces the volume viscosity, dynamic modulus, and temperature coefficient of instantaneous modulus. In melts containing ~40 mol.% alkali, the volume viscosity is within experimental error, so only shear viscosity relaxation is observed. The time relaxation spectrum for those compositions degenerates practically into the only relaxation time (see Figure 5). A similar narrowing of the mechanical α-relaxation times spectrum with increasing R_2O contents in the $R_2O–B_2O_3$ systems was observed by Lomovskoi and Bartenev.[64]

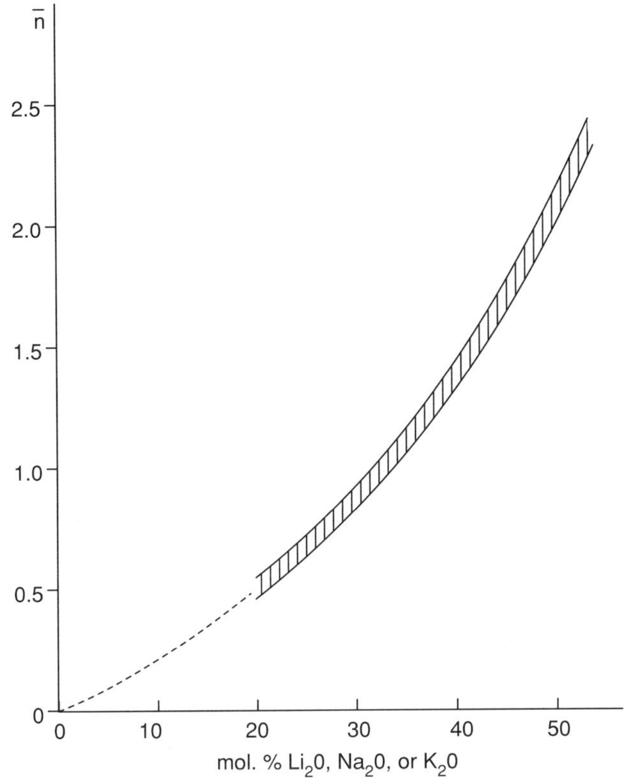

FIGURE 4. The average number of nonbridging oxygen atoms per SiO_4 tetrahedron ($\bar{n}$) in Li_2O–, Na_2O–, and K_2O–SiO_2 systems (generalized data from References 61 and 62).

G. INDEPENDENT PARAMETERS OF STRUCTURE AND THE SILICATE MELT MODEL

In the previous subsection information was presented leading to the following conclusions.[34] Composition regions having a multitude of independent structure parameters ($\Pi > 1$) conform with regions having good glassforming ability. These glasses have structures in which tetrahedra with four and three bridge bonds prevail. Such high spatial connectivity tolerates various distortions of Si–O–Si angles (the entropy rise) for slightly changing volume of the system. This may be understood most distinctly by the example of silica. In vitreous SiO_2, according to X-ray diffraction structural investigations,[65] the Si–O–Si angle distribution is rather wide (from 120 to 180°) with the maximum at 144°. The compressibility as a function of pressure (at 200 K) reaches its maximum (at ~3 kbar) and as a function of temperature has its maximum above zero K — at about 60 K. The thermal expansion coefficient, being negative and close to zero at $T \rightarrow 0$, has a minimum in the neighborhood of 40

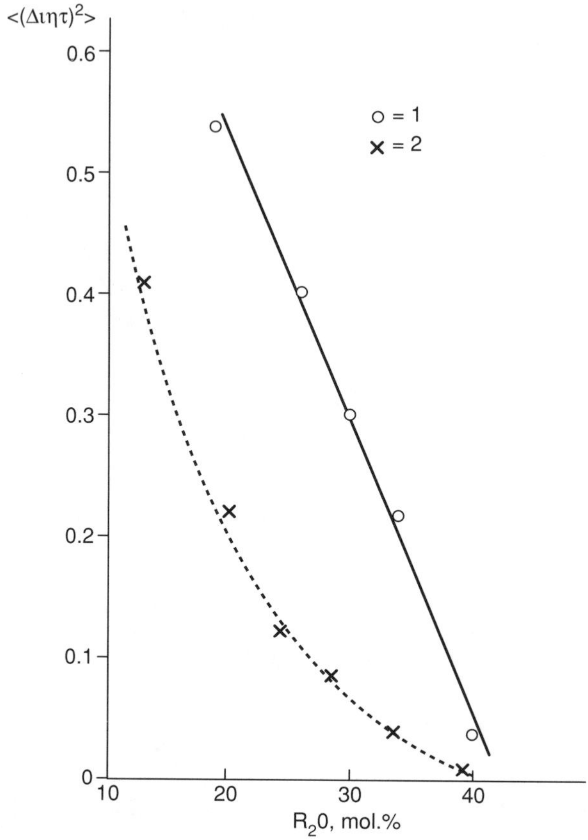

FIGURE 5. Variation of the width of the structural relaxation time spectrum in melts of the Na_2O–SiO_2 (1) and K_2O–SiO_2 (2) systems at 1173 K. (From Nikonov, A. M., Bogdanov, V. N., Nemilov, S. V., Shono, A. A., and Mikhailov, I. G., *Fiz. i Khim. Stekla (Sov. J. Glass Phys. Chem.),* 8, 694, 1982. With permission.)

K and becomes positive above 200 K, its absolute value remaining very low. The heat capacity of the glass below ~60 K is characterized by an excess contribution with respect to Debye's heat capacity calculated for cristobalite. The excess heat capacity is a maximum at about 35 K and vanishes at $T \to 0$ (see review in Reference 66). Vukcevich[66] analyzed the previous models used to explain these anomalies and suggested the most general model. According to his calculations, even for very slight variations of Si–O distances in a tetrahedron (~0.01 Å), the Si–O–Si angle can have two values corresponding to the energy minimum of the system. These angles are comparable with those of β- and α-modifications of crystalline silica, connected by phase transitions. The states of bridging oxygen are separated by a potential barrier, which is due to the dependence of the potential energy on Si–O–Si angle for each stable state and on the Coulomb repulsion of tetrahedra at Si...Si distance changes caused by variations in the Si–O–Si angles (i.e., variation in the distance between tetrahedra centers). The latter term can also be calculated as the energy of elastic strain of the surroundings of a particular SiO_4 tetrahedron when its bridging oxygens are displaced to new positions. The depth of the two potential wells changes with temperature. Therefore, changing temperature varies the population of low-temperature β-states, associated with looser packing, and of α-states, associated with tighter packing. The value of the activation barrier is 4–5 kJ per mole. It is assumed that the spectrum of β- and α-states corresponds to different discrete Si–O–Si angles. Vukcevich's model gives quantitative agreement with the above features of the compressibility and qualitative agreement with the behavior of the thermal expansion coefficient and the heat capacity. The most valuable aspect of this model is the possibility of independent variation of the heat capacity and expansion even for solid glass, based on the specific features of the network structure formed by bridging bonds.

The structural state of such a melt may apparently, in principle, be defined by several independent parameters. One group of parameters can influence the heat capacity and the entropy more but the thermal expansion less. The other group of parameters influences the expansion more and the entropy of the system less.

The ionic interaction increases in going from spatial structures to chains characteristic of metasilicates. $[SiO_2^{2-}O_{2/2}]_n$ chains link to each other by Coulomb forces rather than by van der Waals forces. The former forces act between cations (e.g., Na^+ ions), localized among the chains, and oxygen atoms of the chains. The effective charge of a nonbridging atom is $-|q|$ (where q is the electron charge). According to data from studying positron annihilation, the effective charge of a bridging oxygen atom in vitreous silica is equal to ~–0.62 $|q|$, and in alkali-silicate glasses is a little greater.[67] Therefore, the compensation of the alkali cation charge is achieved by coordination of both nonbridging and bridging oxygen atoms. Varying the position of bridging oxygens by inversion transfers in metasilicate chains with respect to Si...Si lines (as it is for CH_2 groups in polymers) results not only in the increase in chain length and entropy growth, but also in a simultaneous change in the chain packing density. Ionic bonds between chains loosen because of the changing configuration of surroundings (Figure 6). Ionic bonds, unlike covalent ones, are effective at long range. Therefore, any variations of surroundings that involve bridging bonds bearing effective negative charge results in a volume changing. Thus configurational variations even in this rather simplified model appear to be directly connected with volume changes. Apparently, when the melt structure passes over to the metasilicate type, a correlation arises between the parameters defining the entropy (heat capacity) and those defining the volume (thermal expansion). Maybe for that reason Π reduces to 1 (Nemilov et al.[34]). Evidently the above variations of relaxation characteristics correlate well with that qualitative scheme.

Filipovich[68] suggested a semiquantitative version of a similar model using the example of SiO_2 and B_2O_3. This model considers first of all the distortions of bridging valent bonds angles as the cause of disorder appearing in SiO_2. The

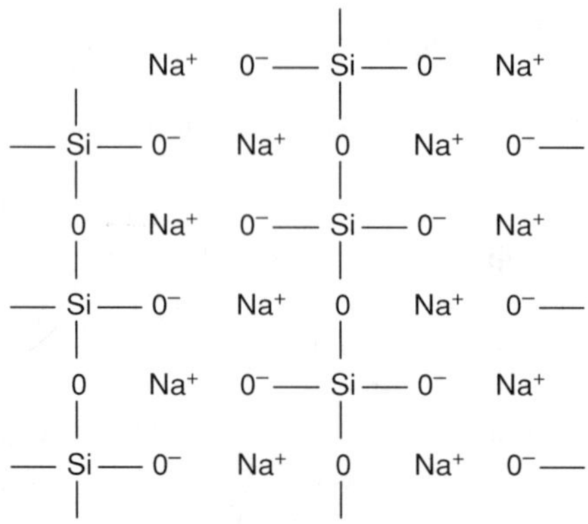

FIGURE 6. Planar scheme of ion coupling for metasilicate chains.

model does not attempt to describe Π changes with variations in composition. Its major result is the explanation of the independence of volume and entropy variations of the network structure.

In the molten SiO_2 structure, the spatial packing becomes less and less tight with increasing temperature owing to the changing orientation of each SiO_4 tetrahedron with respect to the surrounding ones. Quantitatively, such orientation may be simply described by the distribution function ρ of the valent angles β (Si–O–Si). The integral of this distribution function gives the number N of bridging oxygen atoms in the melt, that is,

$$\int \rho(\beta, T) d\beta = N \tag{112}$$

because ρ(β,T)dβ defines the number of bridging oxygen atoms corresponding to the change of Si–O–Si angle from β to β + dβ. At T = 0, $N(\beta_0) = \rho(\beta_0, 0)d\beta_0$. At T > 0, the Gibbs free energy G, the system's configurational entropy S, the number of oxygen atoms n_i to which β angles differing from β_0 correspond, as well as values of the Gibbs free energy φ_i, the internal energy ε_i, the entropy s_i, and the volume v_i associated with each such an atom labeled by i, are defined by the relations

$$G(\beta_0, T) = \sum_{i=0}^{m} n_i \varphi_i - TS \tag{113}$$

$$S = k_B \ln \frac{N(\beta_0)!}{n_1! n_2! \ldots n_m! n_0!}$$

$$n_0 = N(\beta_0) - \sum_{i=0}^{m} n_i \tag{114}$$

and

$$\varphi_i = \varepsilon_i - s_i T + p v_i \tag{115}$$

Such an approach is used to describe two-level systems of point defects in crystals[69] and glasses.[70] Here it is extended to valent angles. Their deviations from the specified β_0 appear as defects.

The energy changes at a valent angle distortion, with respect to its β_0 value, are determined by the equation of nonlinear elastic strain

$$\varepsilon_i = \varepsilon(\beta_0) + \frac{a}{2}(\beta_i - \beta_0)^2 - \frac{b}{3}(\beta_i - \beta_0)^3 \tag{116}$$

where a and b are the coefficients of "harmonic" and "anharmonic" strains respectively. They are assumed to be independent of the absolute value of β_0. From the extremum condition of G ($\delta G = 0$), we have the result $\partial G/\partial n_i = 0$ and $n_i = n_0 \exp(-\Delta\varphi_i/k_B T)$, where $\Delta\varphi_i = \varphi_i - \varphi_0$, i = 1, 2, ..., m. Leaving aside the mathematical transformation procedure, here we give only the expression for the free energy,

$$G(\beta_0, T) = N(\beta_0)\left[\varphi_0 + \sum_{i=0}^{m} x_i\left(\Delta\varphi_i + k_B T \ln x_i\right)\right] \tag{117}$$

where $x_i = n_i/N(\beta_0)$, and for the distribution function

$$\rho(\beta, T) = \int_0^{\pi} \rho(\beta, \beta_0, T)\, d\beta_0, \qquad \rho(\beta, \beta_0, T) = \frac{n(\beta)}{\Delta\beta} \tag{118}$$

where $n(\beta)/\Delta\beta$ is expressed as

$$n(\beta)\frac{d\beta}{\Delta\beta} = \rho(\beta_0, O)\frac{\exp\left(-\Delta\varphi(\beta)/k_B T\right) d\beta/\Delta\beta}{\int_{\beta_0-\Delta\beta_{max}}^{\beta_0+\Delta\beta_{max}} \exp\left(-\Delta\varphi(\beta)/k_B T\right) d\beta/\Delta\beta} \tag{119}$$

$\Delta\beta_{max}$ is the maximum $\beta_0 - \Delta\beta_{max}$ value for the difference modulus $|\beta - \beta_0|$.

Obviously, with increasing temperature, the distribution (118) broadens. Furthermore, Filipovich uses a method of energy calculation for an anharmonic oscillator suggested by Frenkel and obtained for the mean values of the enthalpy fluctuations $\overline{\Delta h} = \overline{\Delta\varepsilon} + p\overline{\Delta v}$, volume $\overline{\Delta v}$, and square volume $\overline{\Delta v^2}$ per unit link the following expressions:

$$\overline{\Delta h} = \frac{k_B T}{2} + \frac{5}{6}\frac{b}{a^3}\left(k_B T\right)^2 \frac{p^2}{2a}\left(\frac{\partial V_1}{\partial \beta}\right)_0^2 \tag{120}$$

$$\overline{\Delta v} = \frac{b}{a^2}\left(\frac{\partial / V_1}{\partial \beta}\right)_0 k_B T - \frac{p}{a}\left(\frac{\partial V_1}{\partial \beta}\right)_0^2 \tag{121}$$

$$\overline{\Delta v^2} = \left(\frac{\partial V_1}{\partial \beta}\right)_0^2 \frac{k_B T}{a} \tag{122}$$

Here the a and b coefficients correspond to those in Equation (116). V_1 is connected with the molar volume of SiO_2 by the relation $V_1 = V/N$; $\Delta V = \overline{N\Delta v}$ (where N is the number of bridging oxygen atoms per mole of melt).

From Relations (120) to (122) the following expressions for molar ΔC_p, $\Delta\kappa_T$, and $\Delta\alpha$ associated only with structure variations of a configurational type are obtained:

$$\Delta C_p = N\,\partial\overline{\Delta h}/\partial T = N\left(\frac{k_B}{2} + \frac{5}{3}\frac{b^2}{a^3}k_B T\right) \tag{123}$$

$$\Delta\kappa_T = -\frac{1}{V_1}\left(\frac{\partial \Delta v}{\partial p}\right) = \frac{1}{V_1}\left(\frac{\partial V_1}{\partial \beta}\right)^2 a^{-1} \tag{124}$$

$$\Delta\alpha = \frac{1}{V_1}\left(\frac{\partial \Delta V_1}{\partial T}\right) = \frac{1}{V_1}\left(\frac{\partial V_1}{\partial \beta}\right) \cdot b \cdot k_B a^{-2} \tag{125}$$

At the glass transition temperature $T = T_g$ the value of $\partial V_1/\partial\beta$ is frozen in and the jump values are fixed. The expected increase of the molar heat capacity at the glass transition point must be close to $Nk_B/2$ value. For silica, the molecule of which includes two oxygen atoms, $N = 2N_A$ (where N_A is Avogadro's number). The experimental value of $C_p \simeq 8\ J \cdot K^{-1} \cdot mol^{-1}$ agrees with the expected one.[71] The configurational heat capacity in this theory appears independent of the volume derivative of the structure parameter β, unlike the compressibility and the thermal expansion coefficient. All three thermodynamic coefficients have different dependencies upon the coefficients of harmonic and anharmonic strain. Therefore, the heat capacity (and the entropy) relaxation must differ from the volume relaxation for thermal or baric variations of the system. For this model the inequalities (72) and $\Pi > 1$ must be true.

At first sight, it seems that the only structure parameter in this theory is the angle β of the bridging bond Si–O–Si. The change of this parameter from the initial to the equilibrium value can be normalized and defined as a general coefficient ξ. The coefficients a and b of "harmonic" and "anharmonic" strains, respectively, remain characteristic of the substance. In this approximation, they do not change for variations of the β angle or of the temperature or pressure. However, such a conception appears to be overly simplified. The cause of the apparent plurality of parameters is the presence of harmonic and anharmonic terms in the elastic strain expression (116).

The condition of thermal equilibrium of the system as a whole eventually superimposes mathematical restrictions on the interrelation of the second derivatives of the free energy G with respect to the variables T, p, and strain tensor components. The latter derivatives appear to be different. That is, in this model, the compressibility depends only upon the harmonic coefficient but not explicitly on temperature, whereas the expansion depends upon both the a and b coefficients, and the heat capacity contains an additive component, independent of either of them. The differences in form of the stress tensor for local strains are based on the functional independence of interatomic potential components of different types upon each other (i.e., those defined by "harmonic" and "anharmonic" coefficients, by the Madelung constant, and others). Obviously, the structural parameters in Filipovich's model really are the stress tensors differing for those very causes. The course of their relaxation must be statistically independent whereas the amplitude of the variations can always be normalized and formalized in the form of several independent structural parameters ξ_k.

In their recent work, Visscher and Logan[72] have considered the stress tensor as a possible order parameter for the liquid-glass transition, and its autocorrelation matrix has been measured in simulations.

It should be noted that in the thermodynamic relaxation theory of depolarized-light scattering due to density fluctuations in viscous liquids (structure of the Rayleigh line), a structural ordering parameter appearing as a tensor was introduced as far back as Leontovich.[73] This defined the measure of the deviation of the axis of asymmetrical molecules from their isotropic distribution. This theory was both developed and confirmed (see references in the work by Quentric[74]).

The value of $\Pi = 1$ in terms of the model under consideration should be expected only if a local interaction becomes isotropic (nondirectional). Then either the anharmonic coefficient b disappears or a correlation arises between the a and b coefficients. It may possibly occur on the formation of metasilicate chain structures with prevailing ionic (isotropic) interaction. Unlike the Roe model,[49] the Filipovich model is continuous with a large number of degrees of freedom.

The above works outline the limits of the attempts made thus far to create a theory involving several independent structure parameters for a glassforming liquid. The real physical picture is apparently rather more complicated. Angell and Rao[75] suggested that configurational states of higher energy can reduce the frequencies of atomic vibrations initiated by the enhancement of anharmonicity. Basing on an analysis of the heat capacity of quenched and annealed glasses near T_g and near zero K, Goldstein[76] came to a similar conclusion. To his mind, more than one half of the ΔC_p value may be associated with those circumstances.

H. EXPERIMENTAL EVIDENCE OF THE INTERNAL CHANGES CHARACTERISTIC OF PASSING THROUGH GLASS TRANSITION TEMPERATURE

Looking ahead, it should be noted that investigations of structural relaxation in glasses confirm the independence of volume relaxation and of enthalpy (entropy) relaxation processes (see Moynihan et al.[77]). Therefore, the question of the thermodynamic criterion (72) or (79) for the $\Pi>1$ inequality conforming with direct experiment should be considered to be clear. However, the question of the nature of the changes in the internal structure by which vitreous (below T_g) and liquid (above T_g) states differ is not yet clear.

At the transition from solid glass to liquid in B_2O_3 ($T_g \simeq 540$ K), GeO_2 ($T_g \simeq 840$ K), and SiO_2 ($T_g \simeq 1450$ K), X-ray studies (see Zarzycki[78,79]) do not reveal any variations in coordination numbers of the structural polyhedra of the central atoms (B, Ge, and Si). In GeO_2 and B_2O_3 this method yields only an increase of bridging bond angles. This is partly understandable because there is no entropy jump at T_g and the glass transition is not one of the first order phase transition. In SiO_2 up to 1773 K, the Si–O–Si valent angles decrease and the structure becomes denser (see Sokol'sky[80]). In the temperature range of 2020–2120 K, the structure of the SiO_2 tetrahedron is conserved but becomes gradually more and more open, like that of cristobalite (Nukui et al.[81]).

Some essential changes in the chemical structure of melts of complex composition might be expected with the temperature variation, because in solid glasses such changes are obviously impossible. However, investigations of binary glasses disprove this assumption. The relative content of silicate anions in lead-silicate glasses obtained by quenching at various temperatures is practically the same (as obtained by gas-liquid chromatography data by Smart and Glasser[82]). Radial distribution functions from the X-ray scattering analysis of sodium germanate glasses indicate only a slight decrease (by ~ 1–5 %) of the six-coordinated germanium fraction with increase of the melt quenching temperature (Kamiya et al.[83]). B^{11} NMR spectroscopy of sodium borate glasses showed similar small changing of four-coordinated boron for strongly quenched and throughly annealed samples (Grimmer and Herms[84]). The same result was obtained by Derno et al.[85] on the same compositions by the method of X-ray scattering. These changes are due to normal temperature variations of the chemical equilibrium constants in liquids.

Interesting results are obtained by the vibrational spectroscopy of glasses and their melts. In the Raman spectra of vitreous silica, as the quenching temperature is raised (i.e., the fictive temperature T_f) from 1170 to 1820 K, only slight frequency shifts of the bending vibrations, $\omega_1 \simeq 450$, $\omega_3 \simeq 700$, ω_4 (TO) ~ 1050, and ω_4(LO) ~ 1200 cm^{-1} are noticeable, whereas ω_4 (LO and TO) decrease with increasing T_f and ω_3 and ω_1 increase, as is required by their correlation with the Si–O–Si angle value, the latter decreasing with increasing

T_f (Greissberger and Galeener[86]). Relaxation times for isothermal stabilization found from positions of bending vibrations around T_g are about 10^2 s, but the bands at 492 and 606 cm^{-1} have constant frequencies and their intensity increases with increasing T_f. These frequencies are related to the spectrum of defect centers formed by dangling bonds (these centers are also produced by ionizing radiation). The heat of formation of these defects, as estimated from the change in their equilibrium concentration (~ 10^{19} to 10^{20} cm^{-3}) on the basis of line intensity analysis, are respectively* 13.5 and 38.5 kJ · mol^{-1}. A similar increase of analogous defect center concentrations in GeO_2 with increasing temperature was determined by Seifert et al.[87] from high-temperature Raman spectroscopy.

For alkali silicate and alumina-alkali silicate glasses, the same investigation has found only general broadening of the vibrational bands (100–1300 cm^{-1}, 290–1620 K). Their relative intensity change corresponds to small variations in the concentrations of $Si_2O_5^{2-}$, SiO_3^{2-}, and SiO_2 types of structural groups due to the usual thermochemical dissociation.

Furukawa and White[88] investigated minutely the Raman spectra of B_2O_3 and alkali borate glasses in the temperature range from room temperature to 1020 K. They concluded that the bands at 805 and 770 cm^{-1} shifted very gradually with increasing temperature and had almost the same temperature coefficient $d(\Delta\omega)/dT$, 0.01 $cm^{-1}K^{-1}$, which was practically unchanged in passing through T_g. The authors associated the frequency decrease mainly with thermal expansion but not with structural regroupings. On the contrary, the widths of these bands ($\Delta\omega$) increase with increasing temperature for $T < T_g$ and have a significantly lower temperature coefficient for $T > T_g$. The $d(\Delta\omega)/dT$ coefficient for both bands is almost the same (~ 0.01–0.02 $cm^{-1}K^{-1}$) below T_g. Above T_g this coefficient increases ~ 1.5 times for the 805 cm^{-1} band and ~ 2.5 times for the 770 cm^{-1} band. The cause of the band broadening, according to the authors, may be the fact that above T_g structural disorder becomes more important and also increases vibrational mode damping due to configurational rearrangements. For all the glasses the intensity ratio (I_{770}/I_{805}) also increases in passing through T_g. Since the 805 cm^{-5} band is usually attributed to symmetric vibrations of boroxol rings, the intensity variation is associated with boroxol ring dissociation and with progressive formation of separate $BO_{3/2}$ groups. The $BO_{3/2}$ groups connect the remaining boroxol rings. This process intensifies with increasing temperature.

According to Walrafen et al.[89–91], who used the band intensity at 805 cm^{-1} normalized to the whole spectra intensity (in the 300–800 cm^{-1} region) an enthalpy variation H° of as much as 23.5 ± 4 kJ mol^{-1} of B_3O_6 was obtained for that process (see the model by Leidecker et al.[48]). This value agrees with the numerical value obtained for the same process by neutron diffraction (Johnson et. al.[92]). Evidently, these variations are associated with chemical equilibria of the polymerization type being frozen in below T_g.

---

* The authors[86] have erroneously named these values "activation energies."

The behavior of low-frequency Raman scattering in the range of 20–100 cm^{-1} is of great interest (see the review by Malinovsky et al.[93]). A low-frequency peak is found in all amorphous materials and glassforming liquids, whereas it is absent in crystals. Its spectral shape is the same for substances of different compositions, but its location depends upon the nature of the substance and its temperature. With increasing temperature (above 100 K), this peak is distorted by quasi-elastic scattering. Separating this peak from the Rayleigh line wing is an independent task, the correctness of which defines the result. Frequency-thermal changes in the intensity of this band are connected with the temperature dependence of the number of vibrational excitations via Bose statistics:

$$I \sim [n(\omega)+1] = [\exp(\hbar\omega / k_B T) - 1]^{-1} + 1 \tag{126}$$

Therefore, the peak is called a "boson" peak. Theoretical models of the boson peak are essentially different, and there is no conventional uniformity of opinions on the matter. Therefore, we refrain from giving any account of them here (see Reference 93). We note only that the spectral shape of the Rayleigh wing is well described as a sum of central and boson peaks for $F(R) = \exp(-R/R_0)$, where $F(R)$ is a spatial function for the correlation of elasto-optic constant fluctuations and R_0 is the correlation radius. On the other hand, the intensity of the band maximum correlates with the mean square of the density fluctuations as $I_n(\omega_{max}) \sim \langle\delta\rho\rangle^2$ and must increase with the increasing T ($\langle\delta\rho\rangle^2 \sim T$). The values of ω_{max}, sound velocity v, and R_0 are expected to correlate as $\omega_{max} \approx (v/\pi)R_0$. Thus, the boson peak behavior is defined by the size effect.

As long ago as 1971, Shuker and Gammon[94] compared Raman spectra in the solid (296 K), softened (550 K), and liquid (1281 K) states of B_2O_3 in the neighborhood of 50 cm^{-1} for Stokes and anti-Stokes shifts. At that time theoretical models for the boson peak were not yet developed. Having calculated the contribution to the scattering intensity component due to the thermal population factor, the authors explained changes in the spectra above T_g by variation of the type of vibrational movements. Notwithstanding the fact that the correlation functions for those movements were not known for sure, the unambiguous conclusion was made that they were short-range correlation functions. In terms of the coherent volume, the variation of type of movement may correlate with the decrease of the mode coherent volume. The probable cause may be increasing anharmonic effects followed by nonlinear changes in the polarizability at atomic displacements. Similar results were obtained later on nitrate glasses.[95]

In similar studies by Walrafen et al.[90,91], it was found that in vitreous B_2O_3, $\omega_{max} = 50$ cm^{-1} is practically constant from 8 to 500 K. When passing through T_g (here $T_g = 553$ K) ω_{max} begins to decrease abruptly, and at 650 K $\omega_{max} = 35$ cm^{-1} (the change has the shape of a diffused "step"). The correlation radius of 10 Å corresponds to the low-temperature region of measurements. It is associated with the movement of whole atomic aggregates, clusters of greater sizes than the boroxol ring size, the latter diameter of which is ~ 5.6 Å. A similar shift of ω_{max} was described by Lorösch et al.[96] but in their case R_0 = 8 Å. The ω_{max} shift was attributed to increasing anharmonicity of vibrations, decreasing force coefficients when passing through T_g with the temperature rise, and increasing free volume.[90] A similar ω_{max} decrease through this region during heating was found by Kawamura et al.[97] for vitreous As_2S_3. In solid glasses (below T_g), depending upon composition, ω_{max} has the maximum value at the exact composition As_2S_3 and falls abruptly with deviations from stoichiometry. In the latter case the connectivity of clusters loosens and the structure loses strength because of chemical causes. Therefore the causes of ω_{max} changes, both for temperature and composition variations, are considered by the authors to be the decrease of coupling between vibrating clusters and their surroundings.

Malinovsky et al.[98] have found that in vitreous glycerol in the T_g region (near 226 K) ω_{max} decreases with increasing temperature (R_0 = 10 Å); i.e., the modulus of the $d\omega_{max}/dT$ derivative grows. In all cases investigated, the boson peak intensity changes in the same manner.

Theoretically, the ratio v/ω_{max} is constant only in a rough approximation (according to the data analysis in Reference 98). According to the theory, v is the hypersound velocity (~10^4 MHz). In glycerol it shows no dispersion in the neighborhood of T_g,[99] the latter being defined as the temperature at $1{\cdot}10^{13}$ P viscosity (relaxation time ~ 10^2 s). For this reason, the linear decrease of v with the temperature rise cannot compensate the stepped decrease of ω_{max} when passing through T_g. At the same time, the constancy of correlational length R_0 when passing through T_g from glass to liquid is also noticed for B_2O_3.[100] The cause of this and the structural sense of R_0 are not yet understood.

Mössbauer scattering differs from X-ray scattering by its high coherence and monochromaticity of the γ-radiation. Chempeney and Sedgwick[101] used the former method for the investigation of organic molecular glasses and crystals in the range from the solid state to the melting points. The temperature dependence of the fraction f of elastically scattered γ-quanta at a given angle exhibits an abrupt fall of f above T_g. Since $f = \beta \exp[-Q \langle x^2\rangle]$, where β is the ratio of Rayleigh scattering to the total scattering (i.e., Rayleigh plus Compton scattering), $Q = (4\pi\sin\theta)/\lambda$, and $\langle x^2\rangle$ is the mean square component of energy shifts, these data definitely indicate vibrational amplitude growth above T_g. The authors have calculated and shown that these vibrations are most probably librations (i.e., hindered rotations or "rocking") of structural groups. Above T_g they transform into free rotations. Diffusional type shifts which define the linewidths become noticeable only above T_g.

Mössbauer investigations in the usual dynamic regime using ^{57}Fe nuclei (ferrocene solutions in glassforming organic solvents) were summarized by Vasquez and Flinn.[102] Below T_g the resonant fraction shows the same behavior as for a crystalline solid. Above T_g the Debye-Waller factor (W) begins to increase steeply because of the appearance of an additional term in its exponent. The Debye-Waller factor describes the change in line shape due to incoherent scattering caused by thermal vibrations. Experimentally it is found from the amplitudes and widths of spectral lines. In these experiments it is not associated with diffusional shifts of molecules and its variations reflect the increase of vibrational amplitudes.

NMR investigations on protons in α-D-glucose pentaacetate (glass and crystal) were done by Ellis and McDonald.[103] It follows from line broadening that at T_g and above new noncorrelated movements of some groups appear. The same conclusion follows from an analysis of the second moments of the line shape. The appearance near T_g of drastic and uncorrelated movement of all atomic units forming the glass was pointed out by Kramer et al.[104] (NMR on ^{19}F, ^{31}P, ^{7}Li in alumina-phosphate, metaphosphate, and borate glasses and melts). A point of interest is that lithium movements begin at lower temperatures than those of phosphorus, fluorine, and boron. It conforms qualitatively with the results in Reference 105, where it is determined that T_g as defined by the flexing point on the ionic conductivity curve is lower than T_g found from thermal expansion.

Also interesting are the conclusions on the dynamics of structural variations in the neighborhood of T_g which were performed by Stebbins[106] (NMR study of ^{29}Si nuclei in alkali and alkali-aluminum silicate glasses and melts). In the glass, lines due to different symmetry groups (e.g., $SiO_{4/2}$ and $SiO^{-}O_{3/2}$) are widely separated in frequency. It appears that at temperatures 200–300°C above T_g, these lines gradually merge into a single narrow line of great amplitude. This is explained by the fact that silicate species in the melt at high temperatures have shorter lifetimes than the NMR time scale (the Larmor resonant frequency is about 35 MHz), but longer than the vibrational time scale. It is assumed that with decreasing viscosity, there is an exchange of silicon atoms between species of different types through the breaking and re-forming of strong Si–O bonds in different groups; i.e., different silicate species actually exchange their Si atoms. In support of this assumption, the author of Reference 106 believe, is the observed increase of spin-lattice relaxation time t_1 for ^{29}Si by almost an order when passing from T_m (melting point) to T_g. Below T_g this parameter is frozen in with a value of $\sim 10^1$ s. Stebbins proposes that above T_g intermediate structures of the high energy defect type appear.

Phillips et al.[107] studied inelastic neutron scattering in vitreous, liquid, and crystalline selenium ($\lambda = 4.1$ Å, $0.3 < q < 2.6$ A^{-1}). They established that the vibrational density of states $g(\omega)$ in liquid selenium in the low-frequency region (below $1 \cdot 10^{12}$ Hz) increased significantly compared to a small decrease of $g(\omega)$ in the range of high frequencies. The shape of the $g(\omega)$ curve becomes convex toward the abscissa axis (ω), which distinguishes, in principle, the liquid state from the vitreous and crystalline states (according to Debye, $g(\omega) \sim \omega^2$). When passing from solid glass to liquid through T_g (303 K), the temperature coefficient of the Debye-Waller factor W increases sharply due to the growing vibrational amplitude and enhancing anharmonicity. The density of states was used to calculate the entropy. It was shown that about one third of the excess entropy in liquids above T_g was due to vibrational entropy. This conclusion is in agreement with the conclusions made by Goldstein.[76] A sharp increase of the temperature coefficient of the mean square atomic displacements above T_g for the same substance in the frequency range above 10^{11}Hz was observed by Buchenau and Zorn[108] (data from neutron time-of-flight measurements). These authors confirmed the data obtained in Reference 106 for the temperature dependence of the Debye-Waller factor.

Saunders[109] studied the baric dependence of elastic moduli C_{11} and C_{44} in selenium on passing through T_g for long-wavelength acoustic phonons (10–1000 MHz). Grüneisen parameters (γ_i) increase insignificantly below T_g, but above T_g their temperature coefficient rises sharply. This feature indicates unambiguously that vibrational anharmonicity increases above the glass transition temperatures.

Optical spectroscopy of d-d transitions in glasses at low concentrations of a d-element oxide show no convincing evidence for internal changes in glasses. At low dopant levels of such oxides to glasses containing alkali, an ion of variable valency can change its coordination with temperature because of the formation of local centers. The chemical arrangement of such centers may differ from the usual glass structure. Only in the case where the variable valency ion is a part of the glass framework can one expect to obtain any information of value from d-d-transition spectra relevant to the question of interest. Nemilov and Mikhailov[110] have studied the spectra of cobalt in vitreous $Co(PO_3)_2$ in the temperature range from 290 to 910 K (at ~10^5 P viscosity). In this substance, cobalt links metaphosphate chains formed by $POO_{2/2}O^-$ units and is surrounded tetrahedrally by nonbridging oxygens (O^-). The data on the intensities and frequencies of absorption bands were treated by Ballhausen-Liehr theory and crystal field theory. It is possible to calculate theoretically the bond ionicity (Q) due to overlapping of atomic orbitals. The constancy of the cobalt coordination number (4) over the whole temperature range was established. The oscillator strength (f), transition frequencies, and bond ionicity vary in such a manner that when passing through T_g (756 K) their temperature coefficients change sharply. With increasing temperature the Co–O bond ionicity increases, whereas the thermal expansion contribution to this change is small, ~1%. The main part of the ionicity increase is due to a change in overlap of the cobalt and oxygen orbitals, which is possible only by the increase of atomic vibrational amplitudes. The cause of this condition is evidently the same as in cases discussed earlier. Similar results were obtained by the same authors[111,112] from investigation of other objects.

A summary of everything stated in this section is presented in Figure 7. At T_g no abrupt changes of the averaged structure are observed because none of the parameters directly connected with the atomic coordination has such a behavior. However, the temperature coefficient of many parameters P connected with structure exhibit such a change. Here we leave aside perfectly clear variations of those characteristics, which are due to freezing of chemical structures at T_g. Their concentration in the melt is defined by the thermodynamic equilibrium (as are polymeric species and defects), and the equilibrium freezing at T_g is trivial.

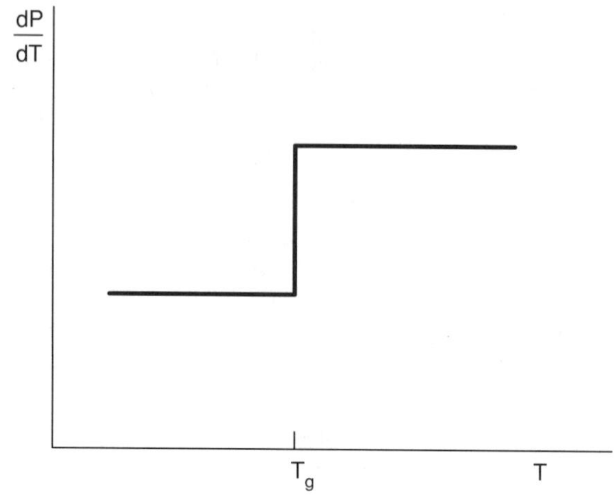

FIGURE 7. General form for the variation of the temperature coefficient of a structure-sensitive parameter P vs. temperature (see text).

Obviously, vibrational spectra line widths in an extended frequency range (though it is yet not shown for the boson peak), NMR line widths, and Mössbauer spectra widths ($\delta\omega \equiv P$) behave as shown in Figure 7. The reciprocal spin-lattice relaxation time ($T_1^{-1} \equiv P$, for NMR), the Grüneisen coefficients for long-wavelength phonons ($\gamma_i \equiv P$), oscillator strength ($-f \equiv P$) for d-d transitions, and the bond ionicity ($Q \equiv P$) vary in the same manner. These variations are due to the same type of change of the Debye-Waller factor ($W \equiv P$). The physical basis for all of these variations is the increase of the vibrational state density in the low-frequency region. The optically active vibrational frequencies of the medium in the spectral range (300–1200 cm^{-1}) are not characterized by a jump of their temperature derivatives at T_g, whereas the boson peak frequency is already displaced, exactly as shown in Figure 7 ($-\omega \equiv P$). Variations of frequencies are caused by the growth of vibrational anharmonicities with increasing temperature and are followed by an increase in vibrational amplitudes. The entire complex system of the mutually correlated parameter changes leads, apparently, to the appearance of new structure configurations and to easing of diffusional-type structure element displacements. The same processes lead to the averaged change of bridging bond angles. However, the width of the latter distribution may vary in a rather peculiar way (e.g., SiO_2), and up to now we have no reliable experimental evidence for the width of the angle distribution varying exactly as shown in Figure 7. At the same time, many hypotheses that in the T_g region configurational changes are followed by "thawing" of free atomic rotations of bridging bonds or whole structural groups seem convincing.

One problem remains obscure, namely, why the arrangement of new structure/configurations requires elementary processes that are typical of viscous flow. We shall come to this issue later when analyzing the kinetic regularities of isothermal transition from glass to liquid; i.e., structural relaxation is limited by energy-consuming kinetic processes, which are responsible for viscous flow. One would think that a configurational variation in the limits of short- and medium-range order (e.g., valent angle) requires no site exchange for structural elements arranging this order. At least all of the material in this section confirms this statement. So far it may be accepted only as a necessity that the whole set of system configurations (the configurational partition function) takes place only when site exchange is possible between the carriers of elementary disorder. In the case of glass these are elementary polyhedra or atoms that form short- and medium-range order around the central atom in a structural polyhedron (bridging atoms). It is, however, doubtless that the very possibility of such atomic jumps over high potential barriers and all the phenomena connected with the atomic vibrations amplitudes, anomalously high in liquids, are closely correlated. This was shown by Buchenau and Zorn[108] for the example of selenium. Its viscosity is reasonably described by the exponential law governing the ratio of the critical mean-square displacement of an atom to the mean-square displacement at a given temperature.

III Thermodynamic Functions of Glasses

A. GENERAL REMARKS

The thermodynamic theory of the glass transition has defined glass as a nonequilibrium system having a free energy excess with respect to metastable liquid. Obviously, glass in that very sense is also a nonequilibrium system with respect to any crystalline modification existing at the same pressure.

In the temperature range above T_g, the liquid metastable state is a function of state and is uniquely defined by the parameters of the state. Therefore, the heat capacity values are quite definite. The heat capacity of glass below T_g is also uniquely defined if, as we have already specified, measurements are done rapidly enough so that structural relaxation is avoided. When rapid relaxation processes occur (e.g., ionic relaxation below T_g), the time scale is always such that these processes are frozen in their own particular temperature range. When approaching T_g, they are completely thawed and do not influence the reproducibility of measurements. Thus, heat can be absorbed and emitted reversibly by a body in both ranges, — i.e., from zero K to T_g and from T_g to the melting point (T_m) — and the problem of irreversible (hysteresis) phenomena may, in principle, be eliminated. The only exclusion is the temperature range ΔT_g which includes T_g when the relaxation time becomes comparable with the measurement time. Jäckle[113] has evaluated the uncertainty in the entropy produced in this interval as

$$\delta S_0 \simeq \frac{1}{8} \Delta C_p\left(T_g\right)\left(\frac{\Delta T_g}{T_g}\right)^2 \tag{127}$$

where $\Delta C_p(T_g)$ is the heat capacity jump and ΔT_g is the actual width of the temperature range where this jump occurs. This uncertainty, for example, corresponds to an entropy evaluation error at zero K of not over 0.2% for glycerol. It is an order of magnitude smaller than those caused by other estimates, and such an uncertainty can be neglected.

The thermodynamic functions of glass, however, are not functions of state but functionals because they depend upon the values of structural parameters frozen in at T_g. Since T_g depends upon the rate at which the system's thermodynamic temperature (or pressure) changes, thermodynamic function values are defined by the history of the resulting vitreous state. However, the relations of the Second Law of Thermodynamics can still be used in calculating the temperature increments of thermodynamic functions because these relations define only the possibility of the system's reversible return or acceptance of heat. The Nernst Heat Theorem (inaccurately named the Third Law of Thermodynamics) will not work, as it establishes that the difference between entropies at 0 K is zero only for internally equilibrium states. For our purposes this condition is not necessary, because the impossibility of glasses being referred to as Nernst bodies does not impose any restrictions on the applicability of the Second Law. To calculate increments of the thermodynamic functions, the condition of $C_p \to 0$ at $T \to 0$ for glass is sufficient.

The above problems have been discussed many times, beginning with the works by Simon.[114] The evolution of these concepts has been reviewed recently by Nemilov,[115] who has suggested a most general formulation of the heat theorem from the viewpoint of irreversible thermodynamics.* The purpose of this chapter is to analyze specific features of thermodynamic functions for glasses as for nonequilibrium systems in the framework of a new, generalized form of the Second Law of Thermodynamics. This form, as will be shown, combines in one expression contributions to changes in thermodynamic functions that are due to both equilibrium and nonequilibrium states.

B. KAUZMANN'S PARADOX

1. The Essence of the Paradox

Even if a relaxational bump of the heat capacity in the T_g region is excluded, the types of change above T_g may differ widely. Figure 8 shows three types of heat capacity variations for a liquid near T_g (a,b,c) in comparison with heat capacity (dotted line, d) for a crystal. The heat capacity of a crystal below T_g usually does not differ much from that of glass. The nature of heat capacity changes cannot be defined as a, b, and c types by any a priori considerations because all three types occur for both inorganic and organic (molecular) substances. Thus far, most important is that in the neighborhood of T_g the value of $\Delta C_p = C_p(l) - C_p(c)$ is large and does not show any apparent tendency to decrease when T_g is shifted very slightly to the low-temperature region. The difference between the entropies of a metastable liquid and of a crystal (S(l) –S(c)) below T_m is determined by the relation

$$S(l)\big|_T - S(c)\big|_T = \Delta S_m - \int_T^{T_m} \left[C_p(l) - C_p(c)\right] \, d\ln T \tag{128}$$

* "The difference in the entropies of any states of a condensed system approching absolute zero is equal to the limit of the sum of the entropy changes at the isothermal irreversible development of internal equilibria of the compared states."

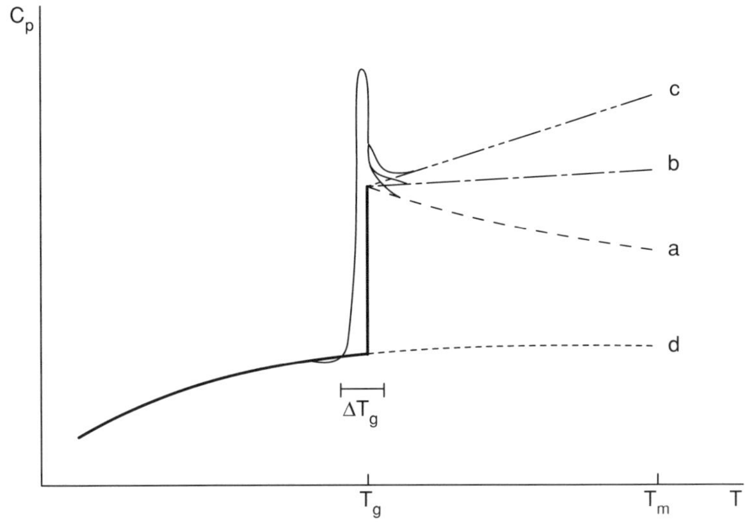

FIGURE 8. Three types of liquid heat capacity above T_g. The thin line stands for the heat capacity course connected with relaxational distortions, a-c (see text).

At $T = T_g$, because of the large difference between heat capacities, the entropy difference between liquid and crystal existing at the melting point (ΔS_m) will be exhausted and for $T_g = T_K$, $S(l) - S(c) = 0$. T_K is called the Kauzmann temperature. Below T_K the entropy of glass would be lower than that of crystal. This situation is shown in Figure 9, where the dotted line below T_g (being defined as usual, e.g., 10^{13} P viscosity) shows the anticipated behavior of the heat capacity (a) and entropy (b) for a liquid. If the heat capacity change and the liquid metastable state at zero K were characterized by the same entropy as that of a crystal, the heat capacity below T_g would have been different (dashed lines in (a) and (b) parts of Figure 9). We know that all of the internally equilibrium states of a substance, including metastable ones, at $T \rightarrow 0$ have indistinguished entropies (the same as for liquid ^{4}He (see Reference 116)).

Obviously a similar paradoxical situation, as in the case of the entropy, will occur with respect to the enthalpy (heat content) and free energy. The latter should be more negative for glass than for crystal; i.e., glass would become more stable than crystal below T'_K. The unnaturalness and paradoxicality of such situation was noticed in the 1930s (see, for example, Blumberg[117]).* He saw a way out of the paradox in the evolution of a molecular aggregation, which he imagined purely qualitatively.

In 1949 Kauzmann[37] illustrated the situation using experimental material including volume variations of systems. Kauzmann saw a way out of the paradox in the fact that, with decreasing T_g, the nucleation of crystals in glass

* "Going ahead and assuming that the heat capacity of glass is more than that of crystal, one can easily make sure from the graph, that at a certain temperature T_0 we would have the true thermodynamic equilibrium between crystalline and glassy phases and below T_0 the glass would have less energy store, i.e. would be a thermodynamically stable phase while the crystal, on the contrary, an unstable one..."[117]

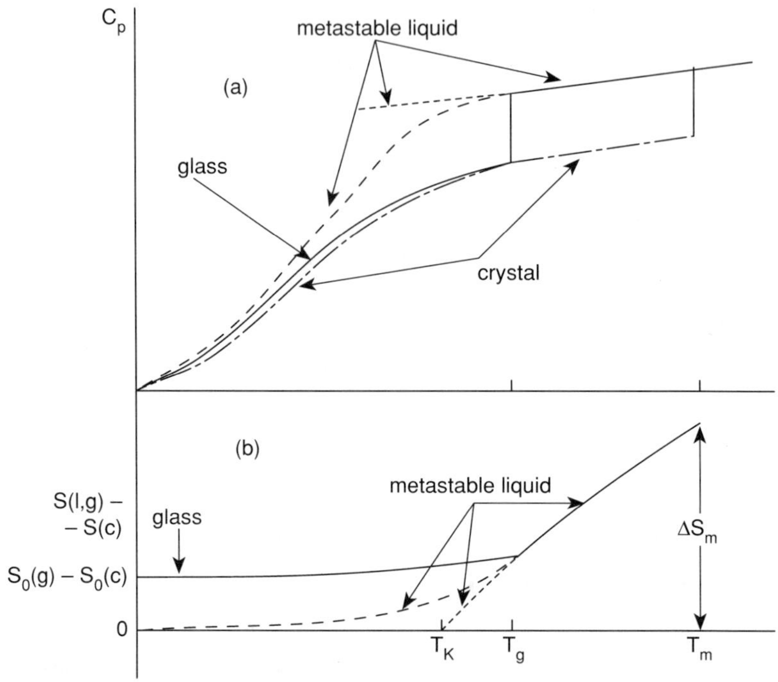

FIGURE 9. Variation in the heat capacity (a) and in the excess, with respect to crystal, entropy (b) for the states of glass and metastable liquid in the temperature range of $0–T_m$. The dotted line shows the heat capacity and excess entropy course, as it would be anticipated if the heat capacity curve were extrapolated from the $T_m–T_g$ region into the region below T_g. The dashed line shows the heat capacity and excess entropy course for the case where the excess entropy of metastable liquid disappears only at $T \rightarrow 0$. T_K is Kauzmann's temperature.

increased, because he assumed the nucleation barrier lowered with approaching T_K. Structural relaxation associated with viscous flow should, in his opinion, be depressed because of growth of the molecular transfer barrier. Kauzmann did not exclude the possibility of a "pseudo-phase" transition at T_K, but he left in doubt the question of the reversibility of such a transition. In the Gibbs and DiMarzio[40–42] glass transition theory, the excess entropy of a polymeric liquid is regarded as a conformation type entropy. At T_K a second-order phase transition is theoretically possible where liquid is the only configuration. That temperature must correspond to the "ideal glass transition."

It should be noted that the Kauzmann Paradox raises a number of rather different questions. Since $T_K < T_g$ (T_g being measured under usual cooling conditions), the problem of the disappearance of excess entropy is automatically associated with that of anticipated extremum behavior of kinetic properties, e.g., of viscosity. The cause of this lies in the fact that the viscosity may be approximated over a fairly wide range of values by the Fulcher-Vogel-Tamman relation (named the FVT Law)

$$\eta = A' \exp\left[\frac{B}{T - T_0}\right] \quad (129)$$

where T_0 is also lower than T_g. In a number of cases (but far from always) it has been successfully shown that $T_0 \approx T_K$. Then the problem of thermodynamic and kinetic features of the systems at $T \to T_K$ or T_0 is to be solved as a united one, whereas the interest in the very nature of ordering below T_g is last. On the other hand, the question of how liquid ordering occurs at T_g is, in itself, independent, bearing its own gnosiological sense. We shall look for the answer to these questions later. We now consider possible ways out of the Kauzmann Paradox only based on facts and some theoretical estimations.

2. Ways out of the Paradox

Long ago, Obald and Newton[118] showed, using the example of glycerol, that annealing in the temperature range a little below T_g results in decreasing S(l) in accord with the dotted line in Figure 9. Gonchukova[119] has done the analysis of the T_g/T_K ratio for glasses of various natures. According to different authors' data, that ratio usually falls within the limits 1.3–1.4, but for the glasses based on salts, it is abnormally low. In particular, for lithium acetate ($LiCH_3COO$), $T_g/T_K = 1.05 \div 1.03$. At $T_g = 413$ K, T_K must be about 393–413 K. Annealing glass in this range (393–413 K), Gonchukova found that the measured heat capacity and the calculated entropy below 413 K changed approximately in the manner expected for a metastable liquid (dashed lines in Figure 9), but not as shown by simple extrapolation from the liquid state (dotted lines). She got temperature dependencies for the completion of structural relaxation and of the beginning of crystallization by homogeneous nucleation which have shown that, in the temperature range investigated, the crystallization time followed that of structural relaxation by more than an order of magnitude.

Similar calculations of the same kinetic dependencies for *o*-terphenyl and anorthite were made by Zelinski et al.[120] According to their data, noticeable crystallization follows by almost two orders, 95% relaxation of the glass entropy. Angell et al.[121] have come to the same conclusion by studying lithium disilicate glass.

DiMarzio[122] has warned against using the relation $S = k_B \ln W$ as the only unambiguous interpretation of entropy because an ensemble of polymeric molecules cannot be considered as a microcanonical one. Then the disappearance of excess entropy at T_K is not to be treated as $W = 1$. In a canonical ensemble, a great number (i) of states of various probability f_i is possible and $S = k_B \Sigma_i f_i \ln f_i$. When approaching T_K from above, in this author's opinion, the number of states reduces first of all because of the impossibility of transition from one configuration to another due to the remoteness of phase space regions conjugated with them. Kerner[123] has explained the paradox in terms of a drastic reduction of the dimension of the phase space and, therefore, the "observable" entropy. Stillinger[124] regarded the problem in the framework of the general formalism of statistic mechanics for metastable states and, especially, for supercooled liquids. In particular, he adheres to the opinion of the Kauzmann Paradox that such a situation is impossible because it supposes the formation of defect-free glass, similar to defect-free crystal. This would require that all the defects would come to the surface and disappear, which is impossible for real interatomic potentials and finite atomic sizes. Besides, the system should have lost its fluidity completely. These conditions are, according to Stillinger, unreal. With decreasing T_g, only the freezing-in of nonequilibrium configurations is possible. Stillinger's theory does not exclude local ordering, but he does not correlate this phenomenon directly with the paradox. We shall return to this model later.

Milchev and Gutzow[125] have considered the configurational entropy of a glassy polymer melt in terms of the lattice–hole model within the framework of the mean field approximation. In this case temperature dependent values are the mean chain-stiffness and the mean polymerization degree. The theory does not assume any anomalies in the neighborhood of T_K because the configurational entropy reduction proceeds gradually. In this connection, we are reminded of a similar simulation of configurational variations in liquid, which uses dynamic connectivity of two states as described by the Schottky function (85).

Murthy[126] proceeded from a qualitative consideration based on the essential divergence (a bifurcation type) between the times of cooperative viscous α-relaxation processes and the times of noncooperative, low-scale β-relaxation processes below a certain temperature T_c (where $T_g < T_c < T_m$ for organic molecular substances). He concluded that progressive clustering of molecules below T_c (this results in increasing cooperativity) produced the necessary entropy minimum. This, however, could not become zero either at $T = T_K$ or at $T \rightarrow 0$ because of the kinetic impossibility of molecules ordering that were not involved into clusters but separated from each other.

Woodcock,[127] having treated the glass transition in terms of the hard spheres model, came to the conclusion that Kauzmann's general idea about ordering is correct. However, it is apparently an additional requirement of a limited thermodynamic transition of the second order associated with the onset of mechanical stability and disappearance of translational diffusion.

Molecular dynamics studies of glass structure lead to concrete conclusions about the character of topological ordering on the scales of short-range and medium-range order. Such conclusions depend on the form of the interaction potentials selected and on the number of particles in the model. Best agreement with experimental data is achieved for SiO_2 (X-ray diffraction, IR spectra, etc.) when the potentials are used that take into consideration the directional character of interatomic bonds and for numbers of particles in the model up to $5 \cdot 10^3$ (Vashishta et al.[128]). Obviously, the solution of the Kauzmann Paradox, from the viewpoint of the possibility structure clusterization will require that the number of particles in the model system be increased by an order of magnitude or two. This is somewhat beyond the actual capabilities of present-day computers.

Summarizing the above discussions, we would note that the problem of finding a way out of the Kauzmann Paradox is only the first problem arising from the attempt to analyze the temperature dependence of thermodynamic functions for glassforming melts. This problem, in my opinion, arises primarily all because of the absence of a liquid heat capacity theory based on a physically substantiated general model. Whereas in crystal theories there are the Born-Karman, Debye, and Einstein theories of heat capacity, there are no such general theories for the liquid heat capacity. It is not clear how thermodynamic functions for metastable liquids should behave at low temperatures. Possible errors are approximately of the same magnitude as if we regarded the Dulong-Petit law valid not only at high but also at medium and low temperatures. Experimental facts support of the supposition that with decreasing T_g, the heat capacity of metastable liquid decreases and becomes less than the value anticipated from its behavior in the high-temperature region.

One probably can't exclude structure ordering of a liquid similar to crystal ordering. The energy of a crystal nucleus of a less than critical size exceeds the energy of liquid only because of the phase-phase separation surface. The absence of the latter of course eliminates using the term "phase," but, apparently, one can't exclude such ordering (without a phase boundary surface) as a way of diminishing entropy and energy. Up to now, it has obviously been impossible to define separately the kinetics of such types of ordering within the total kinetics of structural relaxation.

The second problem, which has not been discussed so far, though it is not connected with the Kauzmann Paradox, is the problem of low-temperature (below T_g) variations of glass heat capacity with decreasing T_g. Indeed, if at T_K the difference between the entropies and free energies of a crystal really disappears, the difference between heat capacities of glass and crystal in the entire temperature range from zero K to T_g must also disappear (or, at least, the difference in the entropy increments in the same region) (Nemilov[129]). We know that in various parts of the 0–T_g interval, the glass heat capacity can be both higher and lower than that of a crystal. Chang et al.[130] have summarized the cases when the glass heat capacity appears to be lower than that of crystal. These are diethyl phthalate, *d*-L-lactic acid, sodium tetraborate, boron trioxide, silicon dioxide, pyroxene, and anorthite. Figure 10 illustrates these data for several glasses graphically in terms of the reduced coordinates. It is important to note that the maximum uncertainty in the experimental determination of heat capacity (δC_p, %) in these works is 3 to 30 times as low as the heat capacity difference detected C_p (g) – C_p(c). For many glasses, negative values of C_p(g) – C_p(c) in the range of 0–T_g are not found, and $C_p(g) > C_p(c)$.

C. HEAT CAPACITY OF GLASSES AND EMPIRICAL CORRELATIONS

1. Quantum Theories of Heat Capacity

The most comprehensive summary of the present situation in heat capacity theory is given in the review by Wunderlich and Baur.[137] Theoretical expressions for the heat capacity of a three-dimensional continuum that may be formed by

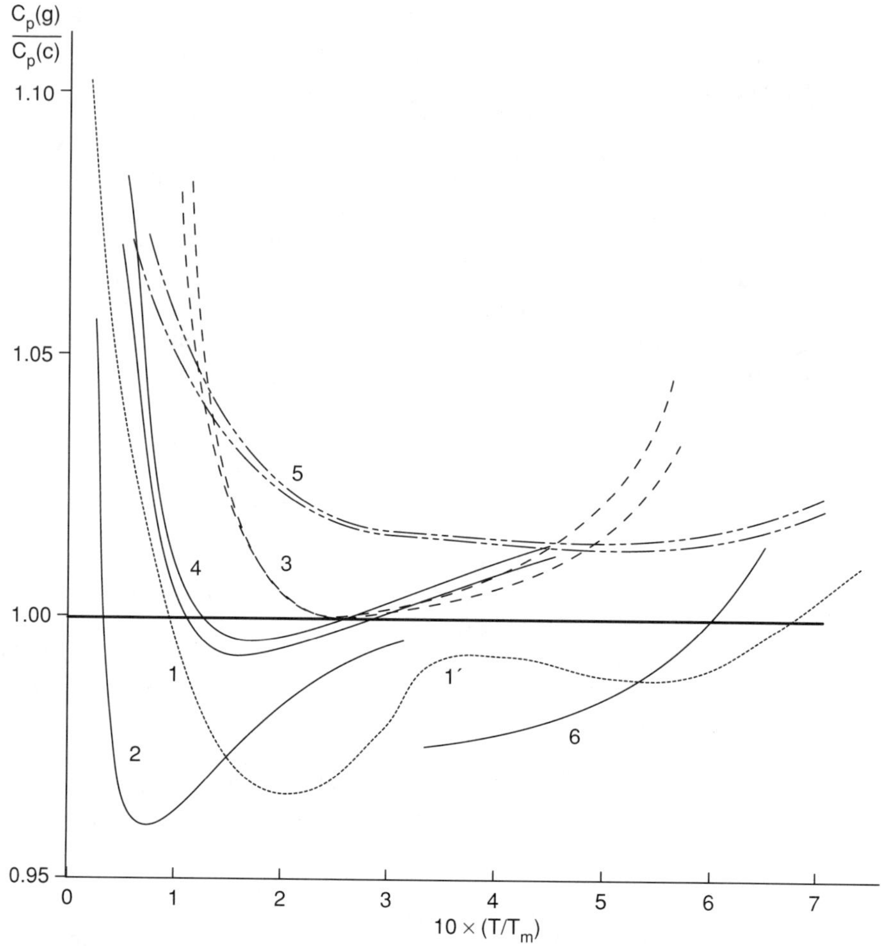

FIGURE 10. Relation of glass heat capacity to that of crystal $[C_p(g)/C_p(c)]$ vs. the reduced temperature T/T_m, where T_m is 1998 K for SiO_2 (β cristobalite), 1015.65 K for $Na_2B_4O_7$, 494.3 K for Se (trigonal), 269.92 K for diethyl phthalate, 329.35 K for *o*-terphenyl, and 289.9 K for dl-Lactic acid. Curves and literature sources are as follows: 1, SiO_2 with respect to α-cristobalite; 1′, SiO_2 with respect to β-cristobalite (see sources quoted by Nemilov in References 131, 132); 2, $Na_2B_4O_7$, Westrum et al.[133]; 3, Se, Chang et al.[134] (upper and lower curves correspond to quenched and annealed glasses, respectively); 4, diethyl phthalate, Chang et al.[130] (the same curve indications); 5, *o*-terphenyl, Chang et al.,[135] (the same curve indications); 6, dl-Lactic acid, Thomas et al.[136] At $T/T_m \rightarrow 0$, $C_p(g)/C_p(c) \rightarrow 1$ (not shown by the graph).

atoms harmonically vibrating near their equilibrium position are defined by extremely simple and well known Einstein and Debye models. Einstein supposed that thermal movement of N atoms situated in lattice sites is equivalent to the movement of 3N independent harmonic oscillators. All frequencies are the same and equal to ω_E. Applying the Planck radiation theory, Einstein[138] found that

$$C_v = 3Nk_B E\left(\frac{\theta_E}{T}\right) \tag{130}$$

where the so-called "Einstein function" is

$$E\left(\frac{\theta_E}{T}\right) = \left(\frac{\theta_E}{T}\right)^2 \exp(\theta_E / T)\left[\exp(\theta_E / T) - 1\right]^{-2} \tag{131}$$

and $\theta_E = \hbar\omega_E/k_B$. The frequency distribution function $g(\omega)$ is extremely simple; it is the Dirac δ-function

$$g(\omega) = 3N\,\delta(\omega - \omega_E) \tag{132}$$

At $T/\theta_E \ll 1$, the heat capacity decreases exponentially with the temperature, whereas at $T/\theta_E \gg 1$, the Dulong-Petit law must hold.

In the Debye model,[139] the solid state is regarded as an isotropic three-dimensional elastic continuum. It is characterized by one branch of longitudinal and two branches of mutually degenerated transverse vibrations for each wave vector. Their propagation velocities (c_l and c_t, respectively) are governed by the relations of the theory of elasticity. The vibrational spectrum is normalized within the frequency continuum from 0 to ω_3 so that

$$\int_0^{\omega_3} g(\omega)\,d\omega = 3N \tag{133}$$

where N is the number of atoms in the system. The number of degrees of freedom Z for a solid of the volume V is given by

$$Z = \frac{V}{6\pi^2}\left(\frac{1}{c_l^3} + \frac{2}{c_t^3}\right)\omega^3 \tag{134}$$

and the square law frequency distribution is

$$g(\omega) = \frac{V}{2\pi^2}\left(\frac{1}{c_l^3} + \frac{2}{c_t^3}\right)\omega^2 \tag{135}$$

The limiting frequency ω_3 is calculated using the velocities of acoustic wave propagation:

$$\omega_3 = \left[\frac{8\pi^2 N}{V} \cdot \frac{c_l^3 c_t^3}{c_t^3 + 3c_l^3}\right]^{1/3} \tag{136}$$

According to the Debye theory, the heat capacity is calculated as

$$C_V = 3Nk_B D_3(\theta_3/T) \tag{137}$$

where the Debye function (for a three-dimensional continuum) is defined by the relation

$$D_3\left(\frac{\theta_3}{T}\right) = 3\left(\frac{T}{\theta_3}\right)^3 \int_0^{\theta_3/T} \left(\frac{\theta_3}{T}\right)^4 \exp\left(\frac{\theta_3}{T}\right)\left[\exp(\theta_3/T) - 1\right]^{-2} d\left(\frac{\theta_3}{T}\right) \tag{138}$$

where $\theta_3 = \hbar\omega_3/k_B$. For $(T/\theta_3) \ll 1$ $C_v \sim (T/\theta_3)^3$ and $(T/\theta_3) \gg 1$, we have the Dulong-Petit law.

The ideas presented here schematically are far from always realized in practice. Ignoring the actual atomic structure of a substance leads to various specifications of the model or to adjusting it to real atomic structures. The elastic continuum properties are defined by the nature of chemical bonds between the atoms forming the real structure, as regarded by crystallochemistry. Spatial three-dimensional structures such as, e.g., silica phases structures, may be contrasted with layered structures of silicates. In the latter, the layers with ionic-covalent bonds (like Si–O–Si) within one layer are connected to other layers by weaker bonds (such as van der Waals ones) or by nondirected Coulomb coupling, as happens, for instance, in disilicates. Further, this series is continued by structures of the metasilicate type where not layers but chains exist with ionic-covalent interactions between them.

In silicates the transition from three-dimensional spatial structures to one-dimensional ones proceeds by introducing a modifier into the glass network (i.e., an alkali or alkali-earth oxide). Such transitions, on the whole, may also be carried out in other ways. In the series of $GeSe_2$–As_2Se_3–Se, the "network-chains" transition is attained by means of valency alteration and by modifying the type of s-p hybridization of the central atom element bonds (Ge, As, and Se). In organic polymer substances, the chain structure is complicated by modifying the repeating link, introducing transverse linkages, etc.

All varieties of spatial structures introduce some complications of the frequency spectrum into the theory of their heat capacity. These reflect changing the degree of the continuum spatial dimensions (i.e., networks–layers–chains), complications of the specific structural units that form a continuum of given dimensions, and specific features of the interactions of layers and chains with each other. A review summarizing this subject was also given by Wunderlich and Baur.[137] The book by Tarassov[140] is devoted to the problem of the heat capacity of inorganic glass.

For one-atom solids, the heat capacities of the three crystallochemical types mentioned above are, in generalized form, described by the Tarassov-Yunitsky approximation[137,141]

$$\begin{aligned} C_v/3R = {} & D_1(\theta_1/T) - (\theta_2/\theta_1)[D_1(\theta_2/T) - D_2(\theta_2/T) \\ & - (\theta_3^2/\theta_1\theta_2)\left[D_2(\theta_3/T) - D_3(\theta_3/T)\right] \end{aligned} \tag{139}$$

where $D_2(\theta_2/T)$ and $D_1(\theta_1/T)$ are the Debye functions for two-dimensional and one-dimensional continua:

$$D_2(\theta_2/T) = 2(T/\theta_2)^2 \int_0^{\theta_2/T} (\theta_2/T)^3 \exp(\theta_2/T)\left[\exp(\theta_2/T) - 1\right]^{-2} d(\theta_2/T) \quad (140)$$

and

$$D_1(\theta_1/T) = (T/\theta_1) \int_0^{\theta_1/T} (\theta_1/T)^2 \exp(\theta_1/T)\left[\exp(\theta_1/T) - 1\right]^{-2} d(\theta_1/T) \quad (141)$$

The first has been tabulated by Tarassov[140] and the second by Wunderlich (see Reference 137). For chains constructed of atoms with different masses, the one-dimensional Debye functions have been tabulated by Deltour and Kartheuser.[142,143] The frequency distribution for the two-dimensional continuum increases linearly up to the maximum frequency ω_2. In the case of a one-dimensional continuum, it is constant in the range from zero to ω_1.

The general number of vibrations, 3N, is divided into components

$$3N = 3N_1 + 3N_2 + 3N_3 \quad (142)$$

according to the existence of different θ_1, θ_2, and θ_3. For flat structures (graphite), $\theta_2 = \theta_1 \neq \theta_3$ and $3N_2 + 3N_3 = 3N$. For chains (selenium), $\theta_3 = \theta_2 \neq \theta_1$ and $3N_1 + 3N_3 = 3N$. For diamond, $\theta_1 = \theta_2 = \theta_3$ and $3N_3 = 3N$.

In the Tarassov theory,[140] the contribution of alkali (or alkali-earth) ions to the heat capacity is defined by the Einstein function with appropriate normalization for the total number of oscillators. Expressions for the heat capacity of polymers may contain, in addition, quantities defined by conformational transitions (Wunderlich[137]). Usually they have the form of a Schottky function since they describe two-level systems. It should be noted that such relations usually work quite well for interpolation. Recently Soga[144] and Hirao and Soga[145] have used them to present experimental data on heat capacity C_v in the range 50–300 K for alkali-silicate, -borate, -germanate, and -phosphate glasses. In comparison with experiment, a very slight difference between C_p and C_v is given by the thermodynamic relation

$$C_p - C_V = \alpha^2 \kappa_T V T \quad (143)$$

where α and κ_T are the thermal expansion coefficient and the isothermal compressibility, respectively. Here the heat capacity is expressed by a simple sum of the two-parameter Debye function and the one-parameter Einstein function taken in proportion to the fractions of network-former and network-modifier gram-atoms (M_1 and M_2)

$$C_V / 3R = M_1 D_{1,3}(\theta_1, \theta_3) + M_2 E(\theta_E) \quad (144)$$

where

$$D_{1,3} = (T/\theta_1) \int_{\theta_3/T}^{\theta_1/T} (\theta/T)^2 \exp(\theta/T)[\exp(\theta/T)-1]^{-2} d(\theta/T)$$

$$+ 3(T/\theta_3)^3 (\theta_3/\theta_1) \int_0^{\theta_3/T} (\theta/T)^4 \exp(\theta/T)[\exp(\theta/T)-1]^{-2} d(\theta/T) \quad (145)$$

The θ_1, and θ_3, and θ_E parameters are to be selected arbitrarily. Variation of θ_1 due to the kind of network-former oxide reveals an obvious correlation with the energetics of structural polyhedra central atoms interacting with oxygen surroundings or with local elasticity moduli expressed in the electrostatic approximation

$$\theta_1 \sim (Z_1 Z_2 / mr^3)^{1/2} \quad (146)$$

where Z_1 and Z_2 are the charges of the central ion (Si, B, Ge, and P) and of the oxygen ion, respectively, m is the reduced mass, and r is the distance between ions.

We omit any discussion of models of glass heat capacity, noting only that such theories thus far do not distinguish crystalline and vitreous states.*

Richet and Bottinga[71] indicated that the heat capacity of silicate glasses below T_g does not change (within the limits of measurement error of about 0.3%) for glass fictive temperature T_f varying in a wide range. However, for selenium, diethyl phthalate, and *o*-terphenyl, Chang and Bestul,[130,134,135] using a higher precision of heat capacity measurements (~0.07%), have found a higher heat capacity for glasses which have been obtained by abrupt melt quenching than that for thoroughly annealed glasses. (See Figure 10).

For thermodynamic calculations in the range of medium and high temperatures, simple approximations such as

$$C_p = a + bT + cT^{-2} + dT^{1/2} \quad (147)$$

where a, b, c, and d are constants, are used. Such expressions may also contain other temperature-dependent terms to fit the experimental data.

With respect to the theory of heat capacity all of the above prove that the main contribution to the heat capacity below the crystal melting point (or glass transition temperature) is given by collective vibrations of the framework and

* *Addition in Proof:* Remember only the very serious work by Lifshitz that proved the generality of $C \sim T^3$ law for extemely low temperatures for all types of structures [Lifshitz, I. M., on the thermal properties of chain and layer structures at low temperatures, *Zh. Exper. i Teor. Fit. (Russ. Y. Experim. and Theoret. Phys.)*, 1952, 22, 475]. This theory explains the existence of $C \sim T^2$ and $C \sim T$ dependences only in the middle temperature regions for layer and chain structures, respectively.

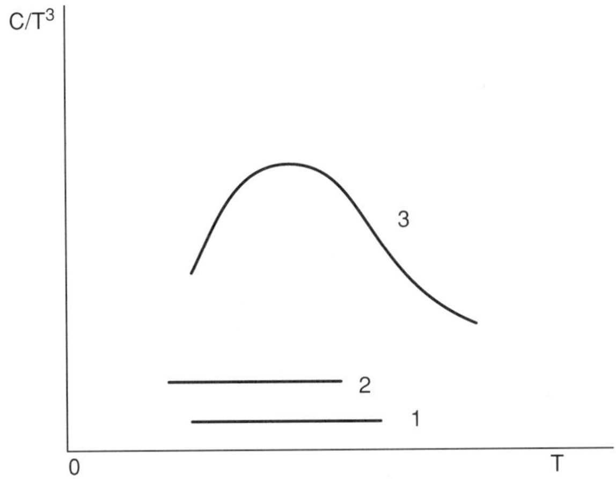

FIGURE 11. Schematic illustration of the deviation of the heat capacity of glass from the Debye law near 0 K. 1, the C/T^3 ratio for crystal; 2, the same ratio for glass based on the elastic constants in the Debye theory; 3, the same ratio for glass, with the experimental heat capacity.

its modifiers; this component of heat capacity may be called a "phonon" one. The term implies a complex vibrational system not always describable by a strict model. The system has a definite vibrational spectrum and this spectrum is specified by the crystallochemical structure, i.e., by definite averaged atomic positions and by force constants of the interatomic bonds.

2. Low-Temperature Anomaly of the Heat Capacity of Glasses

It should be noted that for glasses at low temperatures (lower than $T \sim 10^{-2}\,\theta_D$)* there are significant deviations from the Debye law $C_V \sim T$.[3] The heat capacity of glasses diminishes with the decreasing temperature slower than that of crystals and has a higher heat capacity than crystals of the same composition (Pohl[146]), as shown in Figure 11. A similar anomaly is observed with respect to every property that is sensitive to low-energy excitations.[147] For instance, investigation of the scattering of acoustic phonons shows that Debye phonons in this region are scattered by additional excited states.[147] These may be described by various microscopic models. These phenomena and their theory now represent an independent field of vitreous state physics.

The model of a double-well potential (or a very similar one), the so-called soft mode model, is apparently the basis for the description of the above phenomena. Mutual disposition of the double-well potential minima is defined by the temperature, while the structure of such a defect is formed in the melt. In terms of atomic dynamics, such double-well systems, or soft modes, were

* That is, actually, lower than 5–1 K.

initially introduced only to account for experimental facts, because none of the theories a priori assumed such specific features for a system lacking long-range order. Recently there have been rather promising attempts to follow the genesis of such defects with melt cooling near T_g (Buchenau[148]). The condition of liquid stability requires the absence of stresses. For a high-viscous disordered structure, the vibrational component of the free energy allows formation of unstable regions described by the soft mode dynamics. With increasing temperature, the number of such regions increases. With decreasing temperature, they become localized and stable. The stability is associated with the balance between the vibrational entropy and the energy necessary for their destabilization. These defects are directly connected with the anomalously high density of low-frequency states in glasses and liquids near T_g. This density is due to anharmonicity (see above and the survey by Buchenau[149]). Evaluations of the number of atoms in soft modes based on both excess heat capacity (beyond that calculated by a Debye model) and the criterion of soft mode dynamic stabilization below T_g have led Buchenau[148] to the same numerical value of about 100 atoms.

At low temperatures (~1–10 K), the variation of level populations is determined by tunneling; at high temperatures, the level populations are regulated by the usual mechanism of overcoming the potential barrier. Perhaps this mechanism makes some contribution to low-temperature (below T_g) relaxation.

Soft modes, or soft configurations, are close to unstable ones in a harmonic system, and in solid glass can exist only as localized states. They are characterized by anharmonicity and small force constants for the sole "soft" movement mode. The length of soft configurations is about several interatomic distances. Various model and empirical estimations give the result that in solid glass about 10% of the atoms move in soft configurations. Soft configurations may be associated with the regions of lowered material density, such as voids, or they may be associated with the short-range order distortions in the model of a clustered structure. A detailed review of these concepts is given by Klinger.[150]

For our purposes, it is only essential that strict calculations of glass thermodynamic functions should avoid Debye law extrapolations of heat capacity to very low temperatures. In the 0–50 K region it should be based on measured values. However, the uncertainty in the heat capacity that is caused by using a zero K extrapolation in the 0–10 K region of a power law ($C_p \sim T^{\alpha}$, $1<\alpha<3$) does not introduce appreciable error into calculations. In this temperature region the heat capacity is too low to be able to distort, e.g., the entropy value at zero K, by more than ~0.5%.

3. Empirical Correlations

Approaching the glass transition temperature, the heat capacity of glasses, as a rule, differs only slightly from that calculated by the Dulong-Petit law.

For solid oxide glasses near T_g, $C_p \simeq 3R$ when the heat capacity is attributed equally to all the atoms included in the glass composition (Richet and Bottinga's data[151]). At the glass transition temperatures, the heat capacity of liquid alkali and alkaline earth aluminum silicates and silicates is, on the average, 29 ± 2.9 J · g-at^{-1}K^{-1}. The heat capacity increment ΔC_p is about 4.4 J · g-at^{-1}K^{-1} but varies significantly (more than 3 J·g-at^{-1}K^{-1}) from substance to substance.[151] For borate glasses, ΔC_p varies from 7.6 to 18 J · g-at^{-1}K^{-1}, (Reference 152); for chalcogenide glasses this value lies within about the same limits.[153]

The heat capacity jump at T_g for alcohols, hydrocarbons, monomers, and polymers of the most various compositions shows sufficient constancy ($\Delta C_p = 10.9 \pm 0.8$ J · g-mol^{-1}K^{-1}) that elementary units (beads) repeating in the substructure can be treated as "molecules".* In polyethylene it is the CH_2 group, in alcohols the CH_3, CH_2, CH, and OH groups, etc. (Wunderlich,[154] Wunderlich and Jones,[155] and Varma-Nair et al.[156]). This result was explained[154] in terms of the hole theory by Hirai and Eyring.[36] The ΔC_p constancy for organic substances forms a useful regularity, which should be taken into account.** The absence of a similar strict constancy of ΔC_p for inorganic glasses is thus far not understood. Attempts to ascribe ΔC_p only to an oxygen atom, which might be regarded as similar to a bead in organic glasses, do not diminish the dispersion of ΔC_p.

D. RELATIONSHIP OF THE REDUCED THERMODYNAMIC FUNCTIONS OF VITREOUS SYSTEMS: THERMODYNAMIC INVARIANT OF THE VITREOUS STATE

1. Derivation of the Relation

The vitreous state, as was mentioned above, is characterized by the entropy, free energy, and enthalpy excesses with respect to internal equilibrium phases at the same parameters of state. The problem is to find the general thermodynamic relation between the excess thermodynamic functions of nonequilibrium systems at $T \rightarrow 0$, quasi-reversible increments of thermodynamic functions describing the state changes of nonequilibrium and equilibrium phases in the temperature range from zero K to the phase equilibrium temperature, and the integral degree of glass structure equilibrium at zero temperature.

The De Donder method introduces the degree of conversion (chemical reaction coordinate) ξ, which is normalized so that in the beginning of the

* The molar heat capacity increment is simply divided by the number of beads in a mole of substance.

** Chang and Bestul[135] however, give some examples (cis-1,4 polyisoprene, methanol, selenium, water, isopentane) where ΔC_p is 3 to 4 times smaller than the above-mentioned mean value.

process $\xi = 0$ and at complete conversion $\xi = 1$. Let us select the systems' state at $T \to 0$ as their reference state. Then a structurally nonequilibrium state different from the states of metastable liquids or crystal, must be characterized by nonzero entropy

$$\Delta S_0 = S_0(g) - S_0(msl) = S_0(g) - S_0(c) \tag{148}$$

where $S_0(g)$ is the entropy of a glass, $S_0(msl)$ is that of a metastable liquid, and $S_0(c)$ is that of a crystal, all at zero K. According to the Nernst Heat Theorem, structural equilibrium states at 0 K do not differ in entropy:

$$S_0(msl) = S_0(c) \tag{149}$$

The Gibbs free energies at zero for a metastable liquid and a crystal, unlike the entropies, may be different. The melting entropy determines the maximum possible value of ΔS_0 [see Equation (128)], selecting for a melting point the first one at which glass may be formed at cooling.* At the melting point temperature, T_m, the liquid phase entropy is the function of state. It does not depend upon the way by which the liquid was obtained, whether it was glass heating from zero K or crystal heating with subsequent melting:

$$S_0(g) + \int_0^{T_m} \frac{C_p(g, msl)}{T} dT = S_0(c) + \int_0^{T_m} \frac{C_p(c)}{T} dT + \sum_i \frac{\Delta H_i}{T_i} + \frac{\Delta H_m}{T_m} \tag{150}$$

Here $C_p(g,msl)$ is the heat capacity of glass in the temperature range from 0 to T_g and that of metastable liquid in the range T_g–T_m, ΔH_i are the phase transition heats in the crystal state (i is their number), and T_i are their temperatures ($\Delta S_i = \Delta H_i/T_i$). Melting is specified as $\Delta S_m = \Delta H_m/T_m$. Hence, one can easily calculate ΔS_0 as

$$\Delta S_0 = \Delta S_m - \int_0^{T_m} \left[C_p(g, msl) - C_p(c)\right] d\ln T + \sum_i \Delta S_i \tag{151}$$

The ΔS_0 value does not depend upon whether all of the stable crystalline phases in the range from zero K to the melting point of the highest temperature phase are taken into account or only that high-temperature one is considered that becomes metastable in the low-temperature region. Thus, the degree of structural equilibrium is defined as

$$\xi = 1 - \Delta S_0 / \Delta S_m \tag{152}$$

For the maximum nonequilibrium glass structure, $\xi = 0$; for the utmost equilibrium one, $\xi = 1$.

---

* This assumption is not essential, as it only introduced a conveniently uniform way of formulation.

If at T > 0 structural relaxation occurs, ΔS_0 decreases. The rate of the structural entropy (S_s) production is defined by the relation

$$v = d_i S_s / dt = [d\xi(t)/dt]\Delta S_m \tag{153}$$

The choice of the variable ξ thus corresponds to the most general description of the processes in nonequilibrium systems, because the thermodynamic definition supposes the possible determination of this value on the basis of structural data too.

We now do not consider dynamic regimes and confine our discussion to the thermodynamics of frozen-in nonequilibrium states. The state of liquid at T_m may be characterized by its free energy and enthalpy, which also must not depend upon the way of obtaining the state. Further we shall use the reduced designations of the following form:

$$\Delta S_{0-T_m}(g) = \int_0^{T_m} C_p(g, msl)\, d\ln T \tag{154}$$

$$\Delta S_{0-T_m}(c) = \int_0^{T_m} C_p(c)\, d\ln T \tag{155}$$

$$\Delta S_{0-T_m}(g-c) = \Delta S_{0-T_m}(g) - \Delta S_{0-T_m}(c) \tag{156}$$

$$\Delta G_{0-T_m}(g) = -\int_0^{T_m} dT' \int_0^{T'} C_p(g, msl)\, d\ln T'' \tag{157}$$

$$\Delta G_{0-T_m}(c) = -\int_0^{T_m} dT' \int_0^{T'} C_p(c)\, d\ln T'' \tag{158}$$

$$\Delta G_{0-T_m}(g-c) = \Delta G_{0-T_m}(g) - \Delta G_{0-T_m}(c) \tag{159}$$

$$\Delta H_{0-T_m}(g) = \int_0^{T_m} C_p(g, msl)\, dT \tag{160}$$

$$\Delta H_{0-T_m}(c) = \int_0^{T_m} C_p(c)\, dT \tag{161}$$

$$\Delta H_{0-T_m}(g-c) = \Delta H_{0-T_m}(g) - \Delta H_{0-T_m}(c) \tag{162}$$

For simplicity, we omit here [as well as in Equation (152)], the complication connected with phase transitions in crystalline state.

The two possible ways of obtaining liquid at T_m correspond to the following system of equations:

$$\left.\begin{aligned} S_{T_m}(l) &= S_0(g) + \Delta S_{0-T_m}(g) \\ S_{T_m}(l) &= S_0(c) + \Delta S_{0-T_m}(c) + \Delta S_m \end{aligned}\right\} \tag{163}$$

$$\left.\begin{aligned} G_{T_m}(l) &= G_0(g) + \Delta G_{0-T_m}(g) - T_m S_0(g) \\ G_{T_m}(l) &= G_0(c) + \Delta G_{0-T_m}(c) - T_m S_0(c) \end{aligned}\right\} \tag{164}$$

$$\left.\begin{aligned} H_{T_m}(l) &= H_0(g) + \Delta H_{0-T_m}(g) \\ H_{T_m}(l) &= H_0(c) + \Delta H_{0-T_m}(c) + \Delta H_m \end{aligned}\right\} \tag{165}$$

For excess thermodynamic functions at $T \to 0$ we have, respectively,

$$\Delta S_0(g-c) = \Delta S_m - \Delta S_{0-T_m}(g-c) \tag{166}$$

$$\Delta G_0(g-c) = -\Delta G_{0-T_m}(g-c) + T_m \Delta S_0(g-c) \tag{167}$$

$$\Delta H_0(g-c) = \Delta H_m - \Delta H_{0-T_m}(g-c) \tag{168}$$

We emphasize once more that these relations apply when heat is reversibly accepted/returned by the system. In principle, it is always possible to carry out such a procedure for heat capacity measurements when that condition is fulfilled. The only exception is the glass transition temperature interval (ΔT_g), which we have artificially reduced to one point, having been assured before that the calculational error produced by such a reduction was small enough to be neglected [Relation (127)].

The relations described by Equations (163) to (165) are illustrated in Figure 12. Dashed curves, drawn for comparison, correspond to hypothetical metastable liquid below T_g.

Now we shall find the interrelations of Functions (166) to (168) using the equation

$$\Delta H_0(g-c) = \Delta G_0(g-c) \tag{169}$$

This is used in all thermodynamic calculations for glasses since the work of Nernst and his followers. The condition of $C_p(c) \to 0$ and $C_p(g) \to 0$ at $T \to 0$ is the basis of that equation. We introduce the relation $\Delta G_{0-Tm}(g\text{–}c)/\Delta G_0$

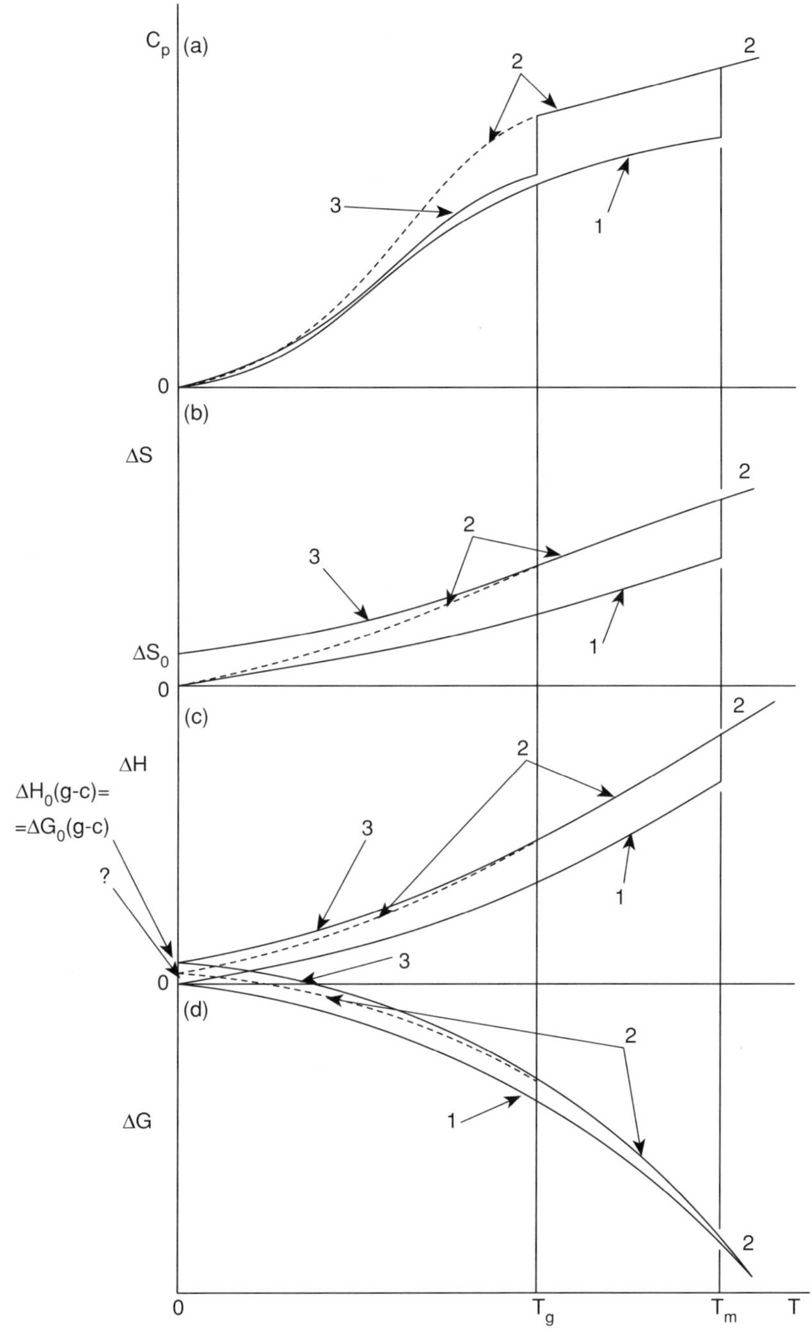

FIGURE 12. Typical course of the heat capacity C_p (a), entropy ΔS (b), enthalpy ΔH (c), and Gibbs free energy ΔG (d) for the states of a crystal (1), an equilibrium liquid (2), a metastable liquid (2, dashed line below T_g, solid line above T_g), and a glass (3). The position marked (?), ΔH_0 (msl – c) = ΔG_0 (msl – c), is thus far unknown. The reference state is the state of crystal at zero K. (From Nemilov, S. V., *Fiz. i Khim. Stekla (Sov. J. Glass Phys. Chem.)*, 7, 575, 1981. With permission.)

(g–c) assuming, at this moment, that ΔG_0 (g–c) $\neq 0$. Adding and subtracting $(\Delta S_0/\Delta S_m)\ \Delta H_{0-Tm}$ (g–c) in the numerator, we use equations (167)–(169) and the relation $\Delta H_m = T_m \Delta S_m$ when grouping the numerator terms. As a result of the transformations applying (152) and (166), we have

$$\frac{\Delta G_{0-T_m}(g-c)}{\Delta G_0(g-c)} = (1-\xi)\frac{\Delta H_{0-T_m}(g-c)}{\Delta H_0(g-c)} - (1-\xi)\frac{\Delta S_{0-T_m}(g-c)}{\Delta S_0} \tag{170}$$

This relation may be called the "thermodynamic invariant of vitreous state" because it includes only the ratios of the function increments and does not include intensive system parameters, such as the molecular weight. The relation (170) defines the correlation between the ratios of those thermodynamic function increments in the interval from zero to T_m, which become real for equilibrium states as reversible or quasi-reversible (for glass) to those function values that at zero K characterize the irreversible transition from glass to crystal. Equation (170) is strict and works almost as precisely as the Second Law of Thermodynamics used in its derivation. The word "almost" implies the maximum error due to irreversible changes in the ΔT_g region, which has already been mentioned. That relation was derived by Nemilov.[129]

2. Analysis of the Relation

Let the designations for individual members of the last relation be introduced as

$$\beta = -\frac{\Delta G_{0-T_m}(g-c)}{\Delta G_0(g-c)} = 1 - \frac{T_m \Delta S_0}{\Delta G_0(g-c)} \tag{171}$$

and

$$b = (1-\xi)\frac{\Delta H_{0-T_m}(g-c)}{\Delta H_0(g-c)} \tag{172}$$

Then after transformation we have

$$\beta + b = \xi \tag{173}$$

The value of β defines the part of the glass excess energy at zero K that is not included in the difference between the heat contents of liquid and of crystal at T_m and that is not connected with the frozen-in part of the melting entropy. It is defined by the difference between the free energies of liquid and crystal that appears at cooling from T_m by all the relaxational phenomena in liquid (the range from T_g to T_m) and by the glass phonon free energy–specific features that distinguish glass from crystal (in the range from 0 to T_g).

The value of b divided by the degree of equilibrium distortion $(1 - \xi)$ determines that part in the excess energy of glass at 0 K which is directly connected with the excess entropy. It is due to the difference in the enthalpies that remains between liquid and glass (on one hand) and crystal (on the other hand) when cooling from T_m for similar relaxational processes in the T_g–T_m interval and by the phonon glass enthalpy–specific features in the range from 0 to T_g.

Relation (173) interrelates the two energetic characteristics of the glass equilibrium degree (β and b) and the degree of its structural equilibrium ξ, the latter being characterized only by its excess entropy. Such a classification may be helpful for finding out how the glass excess energy relaxes at its structure relaxation. If a ΔS_0 decrement results in the rise of ξ, in the limit at $\xi \rightarrow 1$ two cases are possible: (i) $\Delta G_0 \rightarrow 0$ or ΔH_0 (g–c)/ΔH_{0-Tm} (g–c) $\rightarrow 0$, if the limiting state, is undistinguishable from crystal by its energy; (ii) $\Delta G_0 > 0$, if the liquid state at zero K has some excess energy as compared to a crystal.

In case (i) the value of b at $\xi \rightarrow 1$, according to (172), may be any, nondecreasing value because the limit for ΔH_{0-Tm} (g–c) is ΔH_m. In case (ii) b must decrease. The value of β at $\Delta S_0 \rightarrow 0$ is less informative. However, the finite value $\beta < 1$ at $\xi \rightarrow 1$ may indicate that a certain part of the excess energy in liquid during decreasing ΔS_0 is maintained and the ratio $\Delta S_0/\Delta G_0$ also remains constant. Such a situation may occur simultaneously at any infinitesimal but nonzero ΔG_0.

Relation (173) is helpful at solving the Kauzmann Paradox. Indeed, if a $T_g \rightarrow T_K$ $\Delta S_0 \rightarrow 0$, then below T_K the heat capacity difference between glass and crystal must disappear. Otherwise, the Kauzmann Paradox must not appear because below T_K the phase stability would be disturbed again. These circumstances were discussed in more detail by Nemilov.[129]

3. Systematization of Glass Thermodynamic Functions (Thermodynamic Invariant of the Vitreous State)

For calculating all integrals involved in Equation (170) reliable experimental data should be available for heat capacities phase transitions heats, and melting heats of glass, metastable liquid, and crystal modifications in the entire temperature range from the lowest possible temperatures to the high-temperature modification melting point. Substantiation of the choice of data for calculation formed a special task, which required consideration of thermochemical experiments in more detail. In a number of cases the treatment of data in the original work is sufficient so that no additional calculations are needed. We omit here special discussion of experimental data (as well as the listing of numerous references) and refer the interested reader to the papers of Nemilov[131,132,157,158] where such discussions are presented.

Table 1
Calculations of the zero-point thermodynamic functions of glasses, glassy crystals, and polymers and of the parameters ξ, β, and b of Equation (173).

Substance	T_m	T_g	ΔS_m	ΔS_0	$\Delta G_0(g - c)$	ξ	β	b
	K		$J \cdot mol^{-1} \cdot K^{-1}$		$kJ \cdot mol^{-1}$			
Inorganic Glasses								
SiO_2	1998	~1450	5.23	3.81[a]	8.033[a]	0.272	0.052	0.220
			5.23	4.02[a]	7.774[a]	0.231	–0.034	0.265
			4.77	3.35[b]	7.113[b]	0.298	0.060	0.238
			4.08	2.72[b]	5.858[b]	0.333	0.094	0.239
GeO_2	1389	840	12.34	5.36	13.807	0.566	0.464	0.102
B_2O_3	723	540	33.97	11.38[a]	12.376[a]	0.665	0.335	0.330
			33.97	11.09[a]	12.498	0.674	0.359	0.315
BeF_2	825	592	5.77	3.89[a]	3.284[a]	0.326	0.023	0.303
			5.77	3.79[a]	3.393[a]	0.343	0.079	0.264
$Na_2B_4O_7$	1015.65	735	74.66	6.99[b]	21.067[b]	0.906	0.663	0.243
			79.20	11.53[b]	25.690[b]	0.854	0.544	0.310
$CaAl_2Si_2O_8$ (anorthite)	1830	1120	44.27	38.62	62.760	0.128	–0.126	0.254
$NaAlSi_3O_8$ (albite)	1393	1090	44.5	22.4	43.100	0.497	0.308	0.189
$Na_2Si_2O_5$	1147	715	38.66	19.37	30.238	0.499	0.270	0.229
Na_2SiO_3	1361	670	38.32	12.80	7.196	0.666	0.510	0.156
Se, quenched	494.3	304	12.46	3.94	3.873	0.684	0.498	0.186
Se, annealed	494.3	295	12.46	3.59	3.734	0.712	0.524	0.188
As_2S_3	588	450	48.74	20.25	16.736	0.585	0.289	0.296
As_2Se_3	641	443	63.64	35.98	27.991	0.435	0.177	0.258
$H_2SO_4 \cdot 3H_2O$	236.82	153	102.22	24.69	10.293	0.758	0.432	0.326
Organic Molecular Glasses								
Butene-1	87.82	57	43.815	12.59	1.876	0.713	0.413	0.300
Isopentane	113.36	65	45.34	14.06	2.673	0.690	0.404	0.286
2-Methylpentane	119.55	77	52.41	15.27	3.410	0.709	0.465	0.244
Methanol	175.25	103	18.31	7.61	2.326	0.584	0.426	0.158
Ethanol	159	90	31.01	8.93	2.578	0.712	0.449	0.263
Propane-1-ol	148.75	95	36.112	11.34	2.887	0.686	0.416	0.270
2-Methylpropane-1-ol	171.18	118	36.932	9.00	3.195	0.756	0.518	0.238
Glycerol	291	183	62.84	19.41	9.100	0.691	0.379	0.312
Diethyl phthalate								
quenched	269.92	180.8	66.64	22.19	8.719	0.667	0.313	0.354
annealed	269.92	176.5	66.64	19.99	8.138	0.700	0.337	0.363
o-Terphenyl								
quenched	329.35	244	52.196	15.29	8.169	0.707	0.383	0.322
annealed	329.35	242	52.196	15.08	8.031	0.711	0.381	0.330
Isopropylbenzene	177.13	127	41.36	12.00	3.485	0.710	0.390	0.320

Table 1 (continued)
Calculations of the zero-point thermodynamic functions of glasses, glassy crystals, and polymers and of the parameters ξ, β, and b of Equation (173).

Substance	T_m (K)	T_g (K)	ΔS_m (J · mol⁻¹ · K⁻¹)	ΔS_0 (J · mol⁻¹ · K⁻¹)	$\Delta G_0(g - c)$ (kJ · mol⁻¹)	ξ	β	b
Vinyldimethylphenyl-silane	190.7	129	64.4	24.7	6.527	0.62	0.34	0.28
Glassy Crystals								
Ethanol	159	90	31.01	4.24	2.027	0.863	0.667	0.196
2,3-Dimethylbutane	136[c]	75	47.25[c]	7.40	3.202	0.843	0.685	0.158
Cyclohexanol	263.5[c]	145	31.13[c]	5.73[b]	4.427[b]	0.816	0.659	0.157
	265.5[c]	149	33.26[c]	4.73[b]	4.569[b]	0.858	0.725	0.133
Cycloheptanol	258.45[c]	140	3.389[c]	0.699	0.41	0.794	0.567	0.227
Cyclohexene								
I, in relation to stable form, II	138.63[c]	81	30.53[c]	11.67	2.751	0.618	0.412	0.206
III, in relation to stable form, II	169.0[c]	83	19.36[c]	2.61	1.312	0.865	0.665	0.200
Polymers								
Polyethylene	414.6	237	9.913	2.98	2.464	0.699	0.498	0.201
Polytetrahydrofurane	316	186	34.81	15.19	6.301	0.564	0.238	0.326
Transpolypentenamere	293	171	27.57	12.59	5.000	0.543	0.262	0.281
Polyglycolide	502	318	46.9	17.1	1.17	0.63	0.26	0.37
Polydimethylsiloxane	246	151	18.6	8.9	2.70	0.52	0.19	0.33
Poly-ε-caprolactone	342	209	48.00	23.50	10.00	0.510	0.196	0.314
Polyoxacyclobutane	305	195	31.00	7.66	4.853	0.753	0.519	0.234

[a] Calculations for differing values of heat capacity.
[b] Calculations for different ΔS_m values, but for the most recent heat capacity values.
[c] Data are related to the phase transition in crystal state (see text).
From Nemilov, S. V., *Fiz. Khim. Stekla, (Sov. J. Glass Phys. Chem.)* 8, 11, 1982. With permission.

The results of calculations of β and ξ (b = ξ – β), as well as the minimum number of thermodynamic values which may be used for calculating all the necessary integrals, are given in Table 1. Glass transition temperatures are also listed. Table 1, derived from the results in Reference 132 is supplemented by calculations using new data for sodium disilicate and metasilicate from the work by Nemilov and Muratov,[158] and for sodium alumosilicate $NaAlSi_3O_8$ (analbite and glass) from References 71 and 159. Table 1 includes 13 inorganic glasses, 12 low-molecular substances (hydrocarbons, alcohols), 8 polymers, and 6 molecular substances that form "glassy crystals" (crystals with the

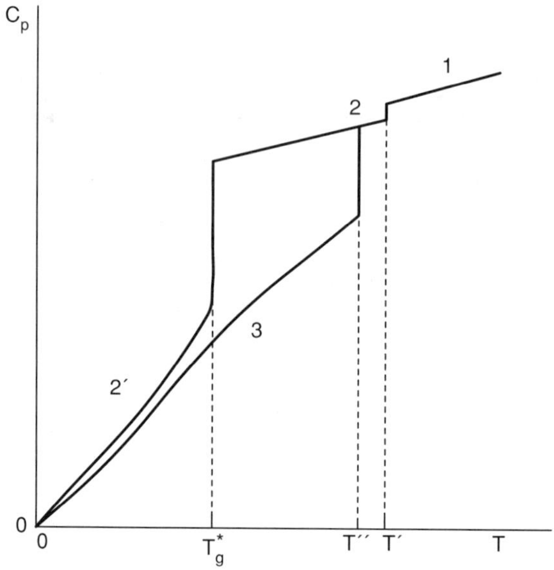

FIGURE 13. Change in heat capacity at the formation of glassy crystal. T′, liquid–crystal I″ phase transition; T″, crystal I–crystal II″ phase transition; heat capacity curve: 1, liquid; 2, crystal I (stable and metastable regions); 2, glassy crystal I below T_g^*; 3, equilibrium crystal II. (From Nemilov, S. V., *Fiz. i Khim. Stekla (Sov. J. Glass Phys. Chem.),* 3, 423, 1977. With permission.)

properties of glass). The latter class of substances is especially interesting for thermodynamics[160] because disordered structure appears in them in addition to a generally regular crystalline lattice. Molecules are situated in specific sites but are chaotically oriented. Such a disordered orientation of molecules usually exists in one of the crystalline modifications, changes with temperature, and gets frozen with cooling at temperature T_g^*, analogous to T_g for ordinary glasses (see Figure 13).* In most calculations related to Equation (170) for glassy crystals, instead of T_m, ΔS_m, and ΔH_m one should use the temperature, entropy, and heat of that phase transition, below which metastable supercooling of the phase appears that causes glass transition. In Figure 13, for example, it is the phase transition at T″. In the case of ethanol, the glassy crystal is formed only at vitrification of the metastable crystallizing phase (127.5 K). All calculations for glassy crystal should be related to this phase melting temperature. Sometimes, a glassy crystal is formed by several metastable crystalline phases. This circumstance is pointed out in Table 1.

It should be noted that all the data on polymers correspond to 100% of the amorphous phase.

The main source of calculational error is usually the melting heat. In this case various versions of the calculations are given. If that circumstance is not

* This class of substances was known in the 1930s but only in 1970s has it got the term "glassy crystals". (For a review, see Reference 160).

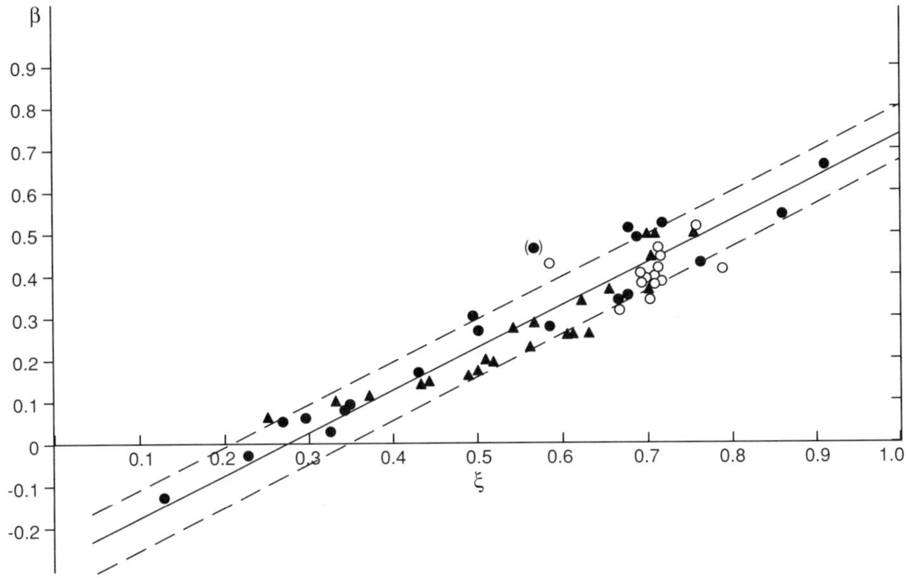

FIGURE 14. Dependence of the β value on the structural equilibrium degree ξ for inorganic glasses (black circles), organic low-molecular glasses (light circles), and 100% amorphous polymers (triangles). In addition to the polymers listed in Table 1, the results of References 156 and 161 to 165 are used. (From Nemilov, S. V., *Fiz. i. Khim. Stekla (Sov. J. Glass Phys. Chem.),* 8, 11, 1982. With permission.)

specified, the second version of the calculation is given for other data on glass or melt heat capacity. The most uncertain species of inorganic glasses is germanium dioxide (in the figures the points are given in parentheses).

Since the publication of Reference 132 much new data on polymers has appeared. These data do not change the results to be given below, so they are not included in Table 1, especially as the tabulated results of thermodynamic calculations for many of them (referring, e.g., to the group of nine alyphatic polyesters[156]) are presented in a form that may very simply be recalculated for the relation (170).

Here we do not discuss the question whether Relation (170) is strictly obeyed (see Nemilov[132]).

The data listed in Table 1 demonstrate that for a wide range of equilibrium glass structure changes (ξ value varying from ~0.13 to ~0.9), the value of b varies within narrow limits. Its mean square value is 0.27 ± 0.06. This is shown in Figures 14 and 15 for three types of glasses: inorganic, molecular, and polymer. The dashed line in Figure 14 defines the range of mean square deviations of b.

The totality of glasses illustrated by Figures 14 and 15 may, with a certain approximation, be regarded as a single glass (i.e., generalized glass) whose degree of structural equilibrium varies in wide limits. We could then consider the dispersion of points as accidental and talk about a sort of idealized experiment

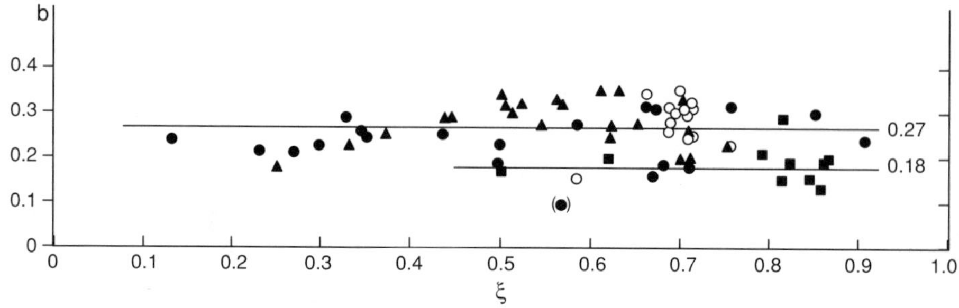

FIGURE 15. Dependence of the value of b [Equation (173)] upon the degree of structural equilibrium ξ for inorganic glasses (black circles), organic low-molecular glasses (light circles), 100% amorphous phase polymers (triangles), and glassy crystals (squares). In addition to the sources mentioned in the caption of Figure 14, the data of Reference 166 are used for pentadecanolactone glassy crystal.

that investigates a generalized glass possessing different equilibra by the parameter ξ. What are the properties of such a generalized glass? First, it is obvious that the parameter b for glasses most probably remains constant or increases only slightly in the relaxation process. That should correspond to the excess free energy of the generalized glass approaching zero at $\xi \to 1$. Such a process would correspond to T_g decreasing ($T_g \to 0$). Thus, the limiting state of metastable liquid at $T_g \to 0$ would be characterized by not only ΔS_0 (msl–c) = 0 but also ΔG_0 (msl–c) = 0 and the thermodynamic difference between liquid and crystal would vanish entirely.

So far the only known example of such a liquid has been ^4He. In this substance for melting $dp_m/dT_m \to 0$ at $T_m \to 0$ and extremely low viscosity favors minimization of the system energy with respect to all ordering sources.[116] At $T_m \to 0$, the difference between the free energies of liquid and crystalline ^4He remains zero. This is never observed for stable and metastable crystalline modifications with the reversible transition temperature above zero. At $T \to 0$, the metastable phase energy is always higher than that of the stable one, though the difference in entropies disappears.

Figure 16 shows the variation of ΔH_0 (g–c)/$\Delta H_{0\text{–}Tm}$ (g–c) = $(1 - \xi)/b$ as a function of ξ for the same substances included in Figure 14. The conclusion about the limiting value of ΔH_0 (or ΔG_0) $\simeq 0$ is apparently not baseless. If these considerations are sound, the Kauzmann Paradox cannot occur at $T > 0$.

These considerations, however, are certainly not free from some sophistication. All the substances studied, being of quite different structure, have been obtained under approximately the same cooling conditions, and the ratio T_g/T_m was, on an average, equal to 2/3.* The very fact of the glasses having such different ξ values is therefore only evidence that in these glasses rapidly and slowly relaxing entropy components exist in various relations. "Configurational entropy" is too general a term, concealing a significant variety of heterogeneous

* The opportunities of theoretical deviations of that very old "2/3 rule" are not to be discussed here. Among a lot of such attempts, I've met no derivation that would not bring in the course of consideration another unobvious premise justifying the result.

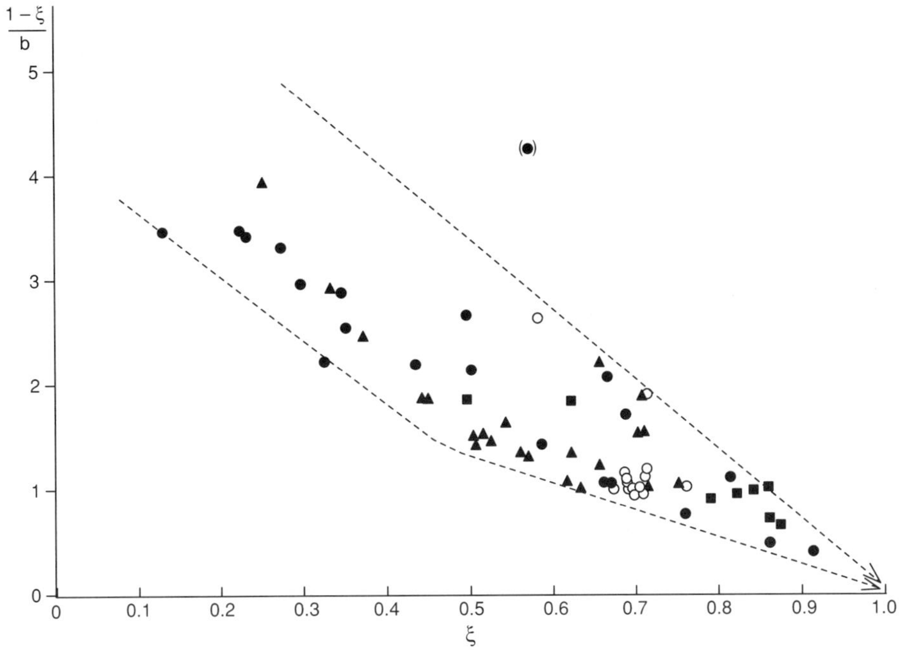

FIGURE 16. Variation of the relation of the residual energy, at zero K, to the 0–T_m enthalpy decrease of liquid (glass), excess with respect to crystal, [ΔH_0 (g – c)/$\Delta H_{0\text{–}Tm}$ (g – c) = (1 – ξ)/b], with the structural equilibrium degree ξ. Here are inorganic glasses (black circles), organic low-molecular glasses (light circles), polymers (triangles), and glassy crystals (squares), with additions listed in the captions of Figures 14 and 15. (From Nemilov, S. V., *Fiz. i Khim. Stekla (Sov. J. Glass Phys. Chem.)*, 8, 11, 1982. With permission.)

factors depending on structure (remember that Goldstein[76] has come to a similar conclusion). That is why those substances that have rapidly relaxing components in small proportion (e.g., SiO_2) have smaller ξ values. Greater ξ values occur for substances in which slowly relaxing structural components associated with the entropy exist in small proportion with respect to rapidly relaxing components (e.g., $Na_2B_4O_7$).

Nevertheless, the totality of various substances with different ξ values might represent a generalized glass if the rapidly and slowly relaxing entropy components could be related to the same regions of a certain generalized phase space. The author of this book is convinced of the possibility of constructing so general a model after having obtained the results described above.

4. Changing of Thermodynamic Functions as a Result of Annealing

Gujrati and Goldstein,[167] by the example of many organic molecular glasses, have shown convincingly that at T_g about 30% of the glass entropy excess, as compared to crystal, is due to the difference between glass and crystal vibrational (phonon) entropy. About 70% of the entropy difference at T_g is connected with the entropy difference near zero K, i.e., with the real configurational contribution. Here our task is to analyze in more detail the changes of such relations resulting from glass stabilization. This was done partly by Nemilov in Reference 132.

Thus far only three substances are known whose heat capacity has been studied by the same authors for a glass obtained by abrupt quenching of melt and for a glass thoroughly annealed at lower temperature. These are selenium,[134] diethyl phthalate,[130] and *o*-terphenyl.[135] The accuracy of the measurements exemplify the maximum capabilities of modern calorimetry (deviations from the average are less than 0.07–0.1%). The heat capacity measurement range (from ~2 K to T_m) is close enough to zero K to minimize the error in the entropy difference, $\Delta S_0(g\text{–}c)$ (not over 0.01 $J{\cdot}mol^{-1}K^{-1}$) and the enthalpy difference $\Delta H_0(g\text{–}c)$ (not over 1 $J{\cdot}mol^{-1}$). It should be noted that the errors introduced into calculations by the error in determining ΔH_m (not over 0.06%) are important only for evaluating absolute values of $\Delta H_0(g\text{–}c)$, the latter being thus determined here with the absolute error of 5 $J{\cdot}mol^{-1}$. However, for our purposes, this inaccuracy is of no importance because for any substance ΔH_m is a constant value and cannot vary in calculations connected with different glass equilibra. The glass transition temperature was defined in the References 130, 134, and 135 graphically by shifting the intersection point of the heat content curves for glasses of different equilibrium with the equilibrium curve for heat content differences of metastable liquid (Figure 17). For comparison, we have calculated the Kauzmann temperatures by the method described (Figure 9). In Table 2 values of T_g are given. T_K, ξ, b, β, and the components of the latter three values, which are produced by the temperature regions from 0 to T_g and from T_g to T_m, are calculated from

$$\xi_1 = \left\{ \int_0^{T_g} \left[C_p(g) - C_p(c) \right] d\ln T \right\} \left(\Delta S_m \right)^{-1} \tag{174}$$

$$\xi_2 = \left\{ \int_{T_g}^{T_m} \left[C_p(msl) - C_p(c) \right] d\ln T \right\} \left(\Delta S_m \right)^{-1} \tag{175}$$

$$b_1 = (1 - \xi) \left\{ \int_0^{T_g} \left[C_p(g) - C_p(c) \right] dT \right\} \left[\Delta H_0(g - c) \right]^{-1} \tag{176}$$

$$b_2 = (1 - \xi) \left\{ \int_{T_g}^{T_m} \left[C_p(msl) - C_p(c) \right] dT \right\} \left[\Delta H_0(g - c) \right]^{-1} \tag{177}$$

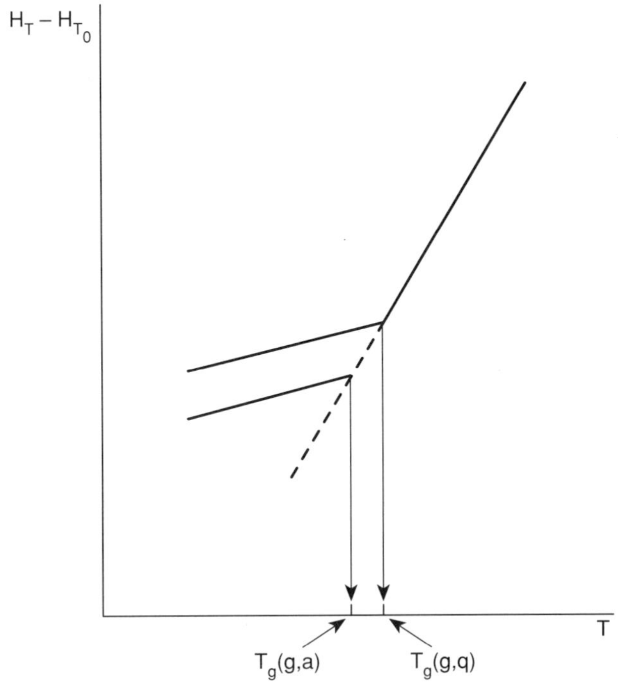

FIGURE 17. Graphical method for determining the glass transition temperatures for quenched (g,q) and annealed (g,a) glasses based on the heat content differences.

$$\beta_1 = \left\{ \int_0^{T_g} dT' \int_0^{T'} \left[C_p(g) - C_p(c) \right] d\ln T'' \right\} \left[\Delta H_0(g-c) \right]^{-1} \tag{178}$$

$$\beta_2 = \left\{ \int_{T_g}^{T_m} dT' \int_0^{T'} \left[C_p(g,msl) - C_p(c) \right] d\ln T'' \right\} \left[\Delta H_0(g-c) \right]^{-1} \tag{179}$$

The part of the phonon entropy, as a component of the entropy difference between glass and crystal at T_g, is calculated using ξ_1 and ξ_2 values, as $\xi_1/(1 - \xi_2)$. It is equal to 43, 16, and 32% for quenched glasses of selenium, diethyl phthalate, and *o*-terphenyl, respectively. These values are close to those found in Reference 167 for other substances.

The reduction of T_g as a result of annealing brings T_g to T_k by 13, 21, and 4% for selenium, diethyl phthalate, and *o*-terphenyl, respectively. The change of the structural equilibrium defined by zero-point entropy (ξ variation with respect to the maximum necessary variation as $1 - \xi$ value), for the same sequence of substances, is respectively 9, 10, and 14%. These values do not correlate with the complete approximation T_g to T_K, though they are of the same order. At the same time, the decrease of $\Delta H_0(g - c)$ upon annealing is

Table 2
Influence of glass annealing on the values of ξ, b, and β [Equation (173)] and their components in the temperature ranges of 0–T_g (index (1)) and T_g–T_m (index(2)).

	Selenium		Diethyl phthalate		o-terphenyl	
	Quenched	Annealed	Quenched	Annealed	Quenched	Annealed
T_K,K	235 ± 10		160 ± 5		193 ± 3	
T_g, K	304	295	180.8	176.5	244	242
$\xi =$	0.684	0.712	0.667	0.700	0.707	0.711
$\xi_1 =$	0.238	0.219	0.066	0.039	0.135	0.123
$\xi_2 =$	0.446	0.493	0.601	0.661	0.572	0.588
$b =$	0.186	0.188	0.354	0.363	0.324	0.330
$b_1 =$	0.013	0.009	0.019	0.012	0.021	0.019
$b_2 =$	0.173	0.179	0.335	0.351	0.303	0.311
$\beta =$	0.498	0.524	0.313	0.337	0.383	0.381
$\beta_1 =$	0.193	0.184	0.038	0.020	0.141	0.128
$\beta_2 =$	0.305	0.340	0.275	0.317	0.242	0.253

only of 3.6, 6.7, and 1.7% of the initial ΔH_0 value in the each case. One may suppose that for the complete shift of T_g to T_K, $\Delta H_0(g - c)$, in general case, would not be zero. The variation of ΔH_0 (g – c) upon annealing with respect to maximum possible difference between the heat contents of liquid and that of crystal can be calculated as

$$\frac{\delta\left[\Delta H_0(g-c)\right]}{\Delta H_m} = \delta\left(\frac{b}{1-\beta}\right) \tag{180}$$

For the same sequence of substances, these values are –2.3, –3.2, and –0.8%, respectively.

Within the framework of the above concept, the increasing b value could be evidence that the limiting state of glass stabilization is the crystalline state, were the region of ξ values close to 1. Here ξ is less than 1, and one cannot be quite sure about such a limiting state based on the analysis of states far from the entropy equilibrium. However, one can assert unambiguously that if the structural processes at annealing are developed in the same manner at T_g approaching T_K, as they have been developed in these experiments, at T_K the glass will have a significant store of the excess zero-point free energy with respect to the crystal.

In the annealing process, in every case of ξ increasing, ξ_1, b_1, and β_1 decrease. This is clear from the scheme of heat capacity change shown in Figure 18. Evidently, movement of the heat capacity "step" towards low temperatures results in the increase of ξ_2, b_2, and β_2 values. Reducing the heat capacity of solid glasses during the annealing process leads to decreasing ξ_1, b_1, and β_1.

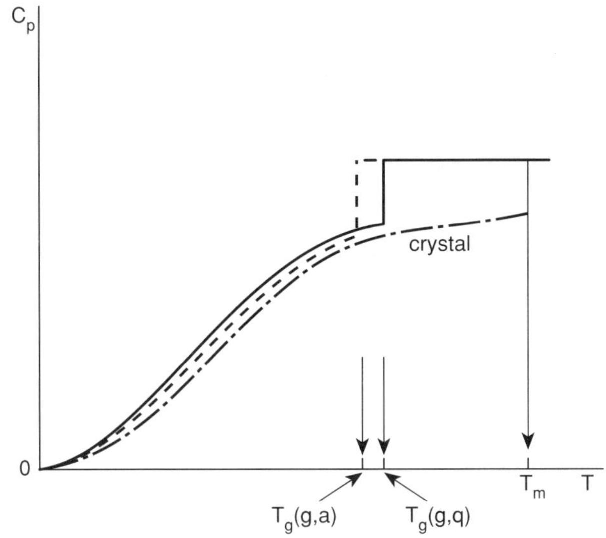

FIGURE 18. Heat capacity changes at the formation of quenched (g,q) and annealed (g,a) glasses.

The tendencies in the difference thermodynamic function variations in both temperature ranges at annealing appear to be contrary. In the region above T_g for liquid cooling, in all cases the processes make the entropy of the states of liquid and crystal more equal. In that sense the more at equilibrium is the state produced by glass transition, the less the subsequent cooling (below T_g) promotes that very tendency of the two states being drawn to each other. In the region above T_g molecular transport processes and thorough structural rearrangements are possible, whereas below T_g they are frozen in. In this region only the phonon entropies, enthalpies, and free energies can promote further ordering of the system on the whole. The capability of the two states being drawn together only by the vibrational movement's energy and entropy appears stronger for quenched glasses than for annealed ones.

5. Nonequilibrium State of Glass as a State of Unbalanced Interaction Potential Energy and Interacting Particles' Heat Energy

The present data may be used for a more profound analysis of the difference between thermodynamic functions of nonequilibrium and equilibrium states. Thus far we have been examining the thermodynamic functions of glass and metastable liquid with respect to the crystalline state. For making a similar analysis of the difference between glass and metastable liquid states, it would be helpful to have metastable liquid thermodynamic functions at low temperatures. However, the latter is impossible, and we are forced to confine ourselves to the comparison of two states, namely, of quenched and annealed glass. A quenched glass (g,q) at T = 0 has an excess in its entropy and free energy, as compared to an annealed glass (g,a), that is,

$$S_0(g,q) - S_0(g,a) = \Delta S_0(g,q-c) - \Delta S_0(g,a-c) \tag{181}$$

and

$$G_0(g,q) - G_0(g,a) = \Delta G_0(g,q-c) - \Delta G_0(g,a-c) \tag{182}$$

The values of the function differences in the right-hand sides of Equations (181) and (182) listed in Table 1 are calculated with respect to the same (crystalline) reference state. Their difference gives the distinction between the functions' absolute values of quenched and annealed states (left-hand sides). Since the free energy and the entropy of each state at the temperature T is defined uniquely within the precision of the constants

$$G_T = G_0 - \int_0^T S_T(T)dT \tag{183}$$

and

$$S_T = S_0 + \int_0^T C_p d\ln T \tag{184}$$

the difference between the free energies of quenched and annealed glass states is

$$\Delta G_T \equiv G_T(g,a) - G_T(g,q) = \left[G_0(g,a) - G_0(g,q)\right] - T\left[S_0(g,a) - S_0(g,q)\right] - \int_0^T dT' \int_0^{T'} \left[C_p(g,a) - C_p(g,q)\right] d\ln T'' \tag{185}$$

At all $T < T_g$ (for annealed glass), $\Delta G_T < 0$, which corresponds to greater thermodynamic stability of the annealed state. In the right-hand side of Equation (185), the quantity defined only by the difference in the potential energies of the particles' interaction $[G_0\ (g,a) - G_0\ (g,q)]$ is negative. The other two quantities $- T\ [S_0\ (g,a) - S_0\ (g,q)]$, and the integral with the "minus" sign — are positive because $S_0\ (g,q) > S_0\ (g,a)$ and the heat capacity of annealed glass is lower than that of quenched glass (see Figures 10 and 18).

As an example, Figure 19 shows the temperature dependence of the difference in the entropy phonon components for annealed and quenched glasses (diethyl phthalate) for a wide temperature range. It was calculated in a simple

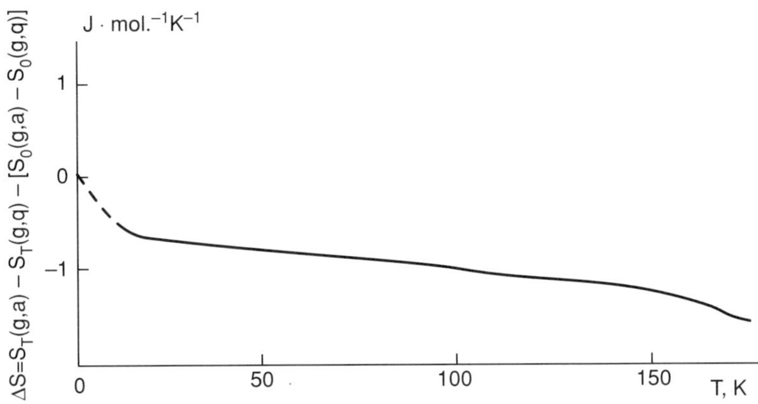

FIGURE 19. Difference between the phonon contribution to the entropy for diethyl phthalate; annealed and quenched glasses.

way using the tables from the work by Chang et al.[130] and taking into account our values for the functions at T = 0. We obtained the same variation course for selenium and *o*-terphenyl.

Figure 20 shows how with increasing temperature the negative term (connected with the interaction potential energy) is compensated by the sum of positive terms (connected with thermal movement of particles). Obviously, when approaching T_g, for quenched glass $\Delta G_T \rightarrow 0$, and the glass vanishes and becomes metastable liquid.*

That calculation convincingly reveals the cause of the unstable state (glass) arising from the metastable liquid. The cause only remotely resembles the distortion of the potential and kinetic energies' equality principle for a system of harmonic oscillators when the oscillators become anharmonic. In our case, the potential energy bears another sense, that is, an excess energy of particles' interaction associated with the disorder (i.e., with configurations). Here the analog of the kinetic energy compensating it is the excess thermal energy. It is defined both by vibrations (the heat capacity contribution) and by excess configurations (the $T\Delta S_0$ contribution). The potential energy of particle interactions in glass is not minimal, and not all the positions of particles (coordinate space) have, as was the case in an equilibrium system, their total equilibrium equivalent for the kinetic energy (i.e., the impulse space region). The divergence in the potential energy, $[G_0\ (g,q) - G_0\ (g,\ a)] > 0$ is due to the fact that interaction of particles in a quenched glass appears weaker than that in a stabilized one. The divergence in the kinetic energy is due to the vibrational state's density in such a system being higher than in annealed glass.

When passing from a particular case of comparing more or less equilibrium states to the general case of comparing internal equilibrium states of metastable liquid and glass, we must provide a scheme similar to that presented in Figure

* For precise calculations one should take into account that in the range of T_g (g,a)–T_g (g,q), the heat capacity of annealed glass transforms into that of liquid, as shown in Figure 18.

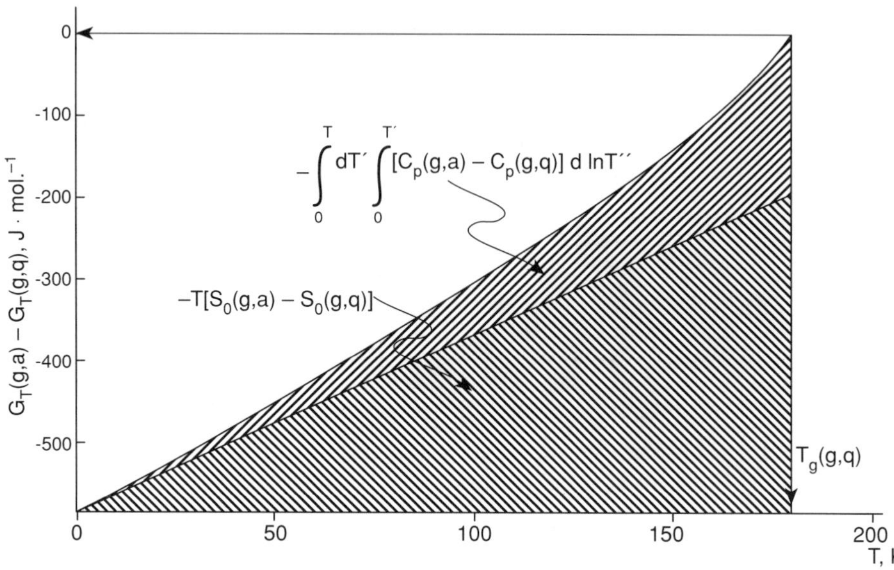

FIGURE 20. Variation of the difference in the free energies of annealed and quenched glasses (diethyl phthalate) associated with the compensation of the "potential" and "kinetic" terms.

20 and replace the designations in Formula (185) so that the entropy phonon component difference should become the difference between metastable liquid and glass entropy phonon components, and ΔG_0 (g) should be counted beginning from the level of metastable liquid — i.e. the (g,a) designation should be replaced by the (msl) designation. At complete internal equilibrium, the excess energy at absolute zero makes no sense (it is zero) and the excess density of vibrational states must vanish as well. However, such a state might arise only at $T_g \rightarrow 0$. Only in this case is the compensation of disappearing ΔG_1 by disappearing positive entropy items possible [see Equation (185)].

Upon cooling a glass, the system on the whole must strive to the equilibrium. Since the average positions of molecules are fixed (ΔS_0 and ΔG_0 are constant), the spectrum of all states of movement must be such that the system state should, as much as possible, approximate the equilibrium at all temperatures. This is realized first of all for low-energy vibrations. Then the appearance of soft modes in the Buchenau model[148] which, as this author believes, obtains dynamic stability with decreasing temperature, also becomes clear. Lasjaunias et al.[168] showed, by the example of B_2O_3, that the excess heat capacity of glass (described by the model of a double-well potential or by the model of soft modes) in the range of 0.1–2 K really increases with the growth of the extent of nonequilibrium (the T_f rise). For other substances, there is a similar heat capacity increase, but in the higher temperature region illustrated by Figure 10.

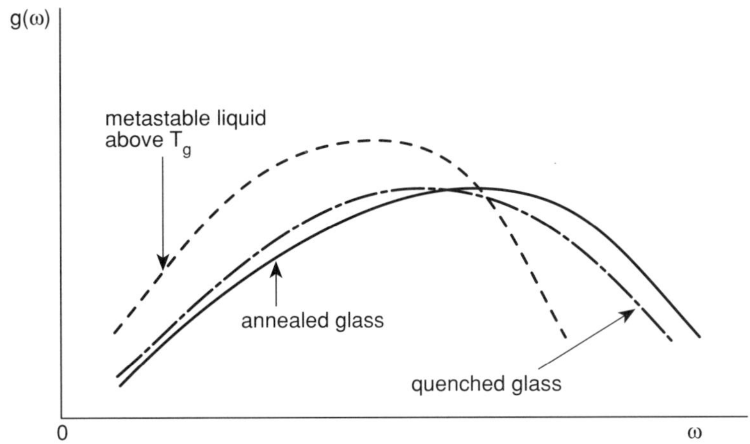

FIGURE 21. Qualitative comparison of the frequency distribution functions for the vibrational spectrum of metastable liquid above T_f, and quenched and annealed glasses (below T_g).

Of course, these considerations are not free from some superfluous determinism with respect to causes and consequential correlations. Splitting the free energy thermal contribution into two members (connected with $T\Delta S_0$ and with the integral), in this sense, is also conventional. It is caused only by the formalism of thermodynamics. There is no doubt that the formation of the vibrational spectrum proceeds already at glass transition and the system of atomic positions itself must initially assume the maximum amplitudes of the free energy kinetic part described above.*

To my mind, a qualitative comparison of the changes in the vibrational states' density versus the degree of glass quenching (Figure 21) makes sense. This scheme follows from the comparison of thermodynamic functions for annealed and quenched glasses, as well as of their heat capacities, which is done in this section. It conforms with the results of neutron inelastic scattering studies in liquid, vitreous, and crystalline selenium (Phillips et al.[107]).

The scheme given here conforms with the considerations of Johari,[169] who also compared quenched and annealed glasses in the low-temperature region (at $0 < T < T_g$). However, he confined himself to the analysis of their entropy and heat capacity, though with extensive involvement of low-temperature (secondary) relaxation results. In quenched glass there is greater heterogeneity in the sense of the presence of a great amount of slackened interaction regions than in annealed glass. Near 0 K these regions produce the excess heat capacity by tunneling and at higher temperatures by realization of states, distinguished also by their vibrational spectrum, through the mechanism of overcoming

* *Addition in Proof:* After the completion of the work on this book, using the method of optimization of thermodynamic functionals depended on ΔG_0 and ΔS_0, the author obtained some new general results (see: Nemilov, S. V., Thermodynamic proof of low temperature excess glass properties inevitability and genetic predestination of dynamic properties of disordered systems at their response to external actions, Sov. J. Glass Phys. Chem., 20, N5, 1994).

barriers (the model of localized double-well potential, low-scale relaxation). Upon annealing a glass, such states become less numerous and its structure, in the sense specified above, becomes more homogeneous. Johari expected that in the limiting case (apparently, when $T_g \to 0$) the glass properties must approach those of an isotropic elastic continuum described by the Debye model. Tunneling effects, together with the thermally activated transitions, gradually cease to dominate.

6. Empirical Structural Correlations of the Zero-Point Entropy

When considering heat capacity we noted that in organic substances a fact of great interest is the approximate constancy of the relation of the heat capacity jump ΔC_p at T_g to the number of beads in a molecule structure. For inorganic glasses bridging atoms act as beads whereas in organic substances elementary structural groups, such as CH_3, CH_2, CH, OH, etc. play the role of beads. By the interaction of such groups (i.e., by van der Waals and dispersive interaction), molecules are links to each other. The structure frozen in is associated with the nonequilibrium orientation of such groups or whole molecules. This is the cause of orientational disorder in the mutual disposition of molecules and a source of excess entropy. Privalko[170] has systematized a large amount of experimental data on vitreous organic polymers and low-molecular organic substances including a small number of inorganic glasses. He established a general empirical correlation between the glass and crystal entropy difference at T_g [ΔS_{Tg} (g–c)] and the entropy of melting (ΔS_m) in form of the relation

$$\Delta S_{T_g}(g - c)/\Delta S_m = M \cdot \exp\left(-N^2 Y^2\right) \tag{186}$$

where $Y = (T_K/T_g) - (T_K/T_m)$ and T_K is Kauzmann temperature. For polymers, $N^2 = 12$ and for other substances, $N^2 = 14$; $M \simeq 1$. Notwithstanding the vagueness of the physical sense of the correlation (186), according to Privalko it indicates the absence of any unified values of ΔS_{Tg} (g–c) or ΔC_p values at T_g, which would have been typical for a linking structural bead.

The most expedient way to look for the relation between structure and the entropy of glasses may be to consider not ΔS_{Tg} (g–c) but only a component of the latter, namely ΔS_0 (g–c). In Table 3 the ΔS_0 (g–c) values per bead, which have been calculated using the data of Table 1 for inorganic glasses, are listed. Evidently, there is no observable constancy of ΔS_0 (g–c)/beads for these glasses. Within an isostructural group of substances, which may be conventionally referred to as "tetrahedral glasses" (SiO_2, GeO_2, BeF_2), relative constancy of ΔS_0 (g–c)/bead does exist.

The data by Varma-Nair et al.[156] for a varied group of linear aliphatic polyesters yields an essential constancy of the ΔS_0 (g–c)/bead values for the group of substances studied: ΔS_0 (g–c)/bead = 4.9 ± 1.5 J $\cdot$ K^{-1}(mol. bead)$^{-1}$

Table 3
Comparison of the excessive, with respect to crystal, zero-point entropies related to a single structural unit (bead), mean values.

Substance	Number of beads in a molecule	$\Delta S_0(g - c)$, $J \cdot K^{-1} \cdot (mol.\ beads)^{-1}$
SiO_2	2	1.7
GeO_2	2	2.68
BeF_2	2	1.92
B_2O_3	3	3.75
$Na_2B_4O_7$	7	1.3
$CaAl_2Si_2O_8$	8	4.80
$NaAlSi_3O_8$	8	2.8
$Na_2Si_2O_5$	3	1.54
Na_2SiO_3	1	3.06
Se	1	3.8
As_2S_3	3	6.7
As_2Se_3	3	12.1
$H_2SO_4 \cdot 3H_2O$	7	3.5

(arithmetical mean). For molecular organic glasses, Nemilov[160] found an approximate constant value of ΔS_0 (g–c)/bead, where partial contributions to the zeroth entropy by CH_x groups (without differentiation between CH_3, CH_2, and CH) and by OH groups were identified, namely

$$\left[\Delta S_0(g-c)/\text{bead}\right]_{\text{partial, for } CH_x} = 2.05 \pm 0.38$$

and

$$\left[\Delta S_0(g-c)/\text{bead}\right]_{\text{partial, for OH}} = 4.81 \pm 0.71\ J \cdot K^{-1}(\text{mol. bead})^{-1}$$

Considering partial contributions given by branches of CH_x groups for glasses containing in a molecule, besides them, 6-membered rings and C = 0 groups (isopropylbenzene has 1 ring + 3 CH_3 groups; *o*-terphenyl has 3 rings, and diethyl phthalate has 1 ring + 4 CH_x groups + 2 C = 0 groups) we can estimate the partial contributions to ΔS_0 (g–c) from a single ring and from a single C = 0 group. These are, respectively 5.5 ± 0.4 and 3.7 ± 1 $J \cdot K^{-1}(\text{mol. bead})^{-1}$. However, they are only estimates because the ΔS_0 (g–c) values depend on the completeness of the glass structure stabilization.* It is interesting to note that Haida et al.,[171] give an almost equal value of Rln2 = 5.8 $J \cdot K^{-1}(\text{mol. ring})^{-1}$ for the disorder entropy at 0 K in cyclohexene, which is due to 6-membered rings at their most disordered arrangement by ring inversion.

* Remember that for molecular organic glasses the ξ-values are nearly the same (see Table 1).

The above numerical results are convincing evidence of inexpediency of a widely spread search for some universal values of ΔC_p, ΔS_0, and other similar differentials to characterize the behavior of a generalized structural bead in glasses of any nature. At the same time, there is no doubt that such simplification makes sense, in principle, when assuming that every bead of the same type within a certain, rather narrow class of substances affects the thermodynamic behavior of glass as discussed in the beginning of this subsection. Such functional individuality of beads enters into the kinetic characteristics of the glass transition (in viscous flow), which are to be considered later.

The absence of ΔS_0 (g–c)/bead constancy is probably due to the fact that the melting entropy ΔS_m includes various components: those due to the disappearance of long-range order, those due to coordination changes or to changes of mutual molecular orientation and of the system topology, those due to variations of the vibrational spectrum, the free volume, the number of defects, etc. These components of the whole disorder, for different groups of substances, may be in different relationships to each other; hence, the contributions ΔS_0(g–c) from different structural groups may differ sharply from each other.

7. Peculiarities of Glassy Crystals: Components of Disorder Entropy

In glassy crystals, there is crystalline periodicity in the arrangement of the centers of gravity of the molecules, but there is no regular orientation of the molecules with respect to each other. This circumstance suggests comparing qualitatively the components of the disorder entropy in glasses at zero K that are due to the absence of long-range periodicity of structure and to the absence of any order in mutual molecular orientation separately. For ethanol (see Table 1), the glass obtained by melt supercooling is characterized by ΔS_0(g–c) = 8.93 $J \cdot mol^{-1}K^{-1}$ and T_g = 90 K. A glassy crystal obtained by a "glass transition" of a crystal crystallizing in a metastable way and having disordered molecular orientation has, at zero K, an entropy of 4.24 $J \cdot mol^{-1}K^{-1}$ with respect to a crystal with regular molecular orientation, and the same T_g value (see Haida et al.[172]). Since both differences (ΔS_0) are measured from the same state (stable crystal), the fact that the glass entropy exceeds that of glassy crystal by 4.69 $J \cdot mol^{-1}K^{-1}$ might be attributed to the disappearance of long-range order. In this substance, the disappearance of long-range order would provide by itself about 50% of excessive zeroth entropy. However, the absence of long-range periodicity has a smaller effect on the ΔG_0(g–c) value. That contribution is only about 20% of ΔG_0(g–c) (see Table 1).

Cyclohexene has two modifications of a glassy crystal resulting from glass transitions of different supercooled crystalline modifications. The entropies of these glassy crystals differ from each other at zero K rather significantly (see Table 1) and points out the different degrees of molecular orientational disorder permitted by the structure of different elementary cells. In a cyclohexene molecule the 6-membered ring is not symmetrical because of the presence of one double bond. Haida et al.[171] believe that in glassy crystal I the degree of disorder of ring orientation is higher due to the possibility of their rotation around the Z axis, which is perpendicular to the plane of molecules, and to the inversion of rings. In glassy crystal III this is not possible and the disorder is caused only by the ring inversion. The latter, however, does not occur completely (the evaluated maximum for this contribution is $R\ln 2 = 5.8\ J \cdot K^{-1}mol^{-1}$).

Orientational disorder in glassy crystals, on the whole, should not always be identified with that in glasses. It must be less because of the limitations superimposed by the existence of the long-range periodicity. For this reason, the contribution to the zeroth entropy of glasses due to the absence of the long-range order must be much less than 50%. For atomic liquids, it is undoubtedly zero (e.g., liquid ^4He).

For the glassy crystals listed in Table 1, one can't calculate precisely the partial contributions to the zeroth entropy from different structural groups. However, rough estimates give values for the contributions to ΔS_0 from CH_x and OH groups of ≤ 1.13 and $1.9\ J \cdot K^{-1}(mol.\ bead)^{-1}$, respectively. They are less than the analogous contributions for normal glasses.

With regard to the general criterion of the degree of structural equilibrium ξ, glassy crystals are closest to equilibrium (see Table 1 and Figures 15 and 16). The value of the b parameter in Equation (173) for glassy crystals is, on average, 0.18 (Table 1 and Figure 15), which is somewhat lower than the mean value for normal glasses (0.27). We shall not comment upon this circumstance. In this case all ξ, β and b values are calculated in relation to a crystal phase transition.

E. GLASS AS A SYSTEM WITH BROKEN ERGODICITY

1. Specific Features of Nonergodicity in Glasses

The excess entropy and the excess free energy existing in glasses with respect to a stable state and to any metastable states raises the question of the inapplicability of some general principles of a statistical description of equilibrium states to the vitreous state. The thermodynamic description of the vitreous state as a nonequilibrium one in the sense specified above does not assume any model concepts about the nature of the glass transition process. From this point of view, the nonequilibrium state of glass is an independent experimental fact.*

The statistical interpretation of the entropy of a system implies, according to Boltzmann, its proportionality to the logarithm of the number of states of the system and, at $T \rightarrow 0$, to the logarithm of the number (W) of configurations that can take place:

* Calculations of the excess thermodynamic functions were carried out in the 1920s when there were no model theories for glass transition (see Simon[173]).

$$S = k_B \ln W \tag{187}$$

In glass, evidently, $S > 0$* because the structure is frozen in and the number of configurations W at all $T \to 0$ cannot vary, only a certain partial set of configurations is possible. From this viewpoint, the entropy of glass cannot be a function of state because the number of configurations corresponds to those realized for an equilibrium state of liquid at T_g (Jäckle,[174] Stishov[175]).

In the phase space of particle coordinates and momenta at specified energy, a certain volume, the value of which defines the entropy, corresponds to the system or to its part (phase). Within this volume, the system particles can have a certain set of coordinates and momenta, all the cells being equally accessible. This means that on the average during a long period ($t \to \infty$) each particle of the system will be in every cell of the phase space and the distribution law does not change with time. The mean for the particle ensemble at any given moment is equal to the mean for a single particle for an infinitely long time. This is the well-known condition of ergodicity. Maintaining internal equilibrium of the system implies keeping this principle. The condition of reversibility of states implies that a new state of the system is also ergodic. A new energy value corresponds to a new value of the phase space volume.

Structural freezing at T_g ($d\xi = 0$) transforms the system from an ergodic to a nonergodic one (Jäckle,[174] Palmer,[176] Stishov,[175] Rostiashvili et al.,[177] and others). This is obvious because not all atomic coordinates (i.e., configurations) are no longer attainable. An ensemble of particles is observed at a given moment (at any moment, glass has a definite structure), but as time $t \to \infty$ there may in principle be structural relaxation in which new configurations appear and former ones disappear (particle coordinates change). Accordingly, the free energy of the system and its entropy also change. Many studies have been devoted to these problems (Jäckle,[174] Palmer,[176] Rostiashvili et al.,[177] and others).

In glass, fragmentary (i.e., incomplete) occupation of coordinate regions in the phase space, in comparison with the equilibrium state, is obviously supplemented by incomplete use of corresponding momentum regions of the phase space. When approaching equilibrium isothermally and at $T_g \to 0$, changes of position and momentum proceed in such way that, simultaneously with the approach to equilibrium, all modes of movement are realized with utmost completeness.

2. Basic Models Describing Glass Nonergodicity

The picture described here is based on experiment and is very rough. It omits some essential details. The very appearance of nonergodicity may be described in terms of distorted statistical symmetry of a system (see, e.g., Forster,[178] Thirumalai et al.,[179] and others). Distortion of lattice symmetry upon crystal melting is described as topological changing of the network, and the

* Strictly speaking, $S = 0$ only for a physical vacuum.

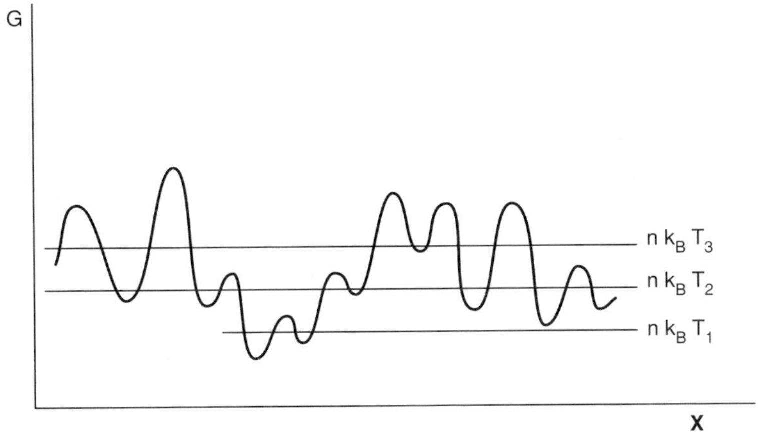

FIGURE 22. Relief of a generalized surface of free energy G vs. generalized coordinates X. Temperatures $T_3 > T_2 > T_1$; n is a constant number (according to ideas of Goldstein,[181] Palmer,[176,180] Stillinger,[182] and others).

disorder entropy in glass may be calculated from a topological model. This question is considered in the next chapter. Here we note only that the opportunity to calculate the excess entropy would bring us only partly nearer to the clarification of the problem. The excess entropy and energetical nonequilibrium states are associated.

Consider the free energy of a dynamic system of particles as a function of their coordinates (Palmer[176,180]). In a generalized coordinates space, very small local regions exist where the neighboring positions distributed at different depths happen to be separated by both small and large potential barriers. The distribution of barrier heights must obviously be continuous (see Figure 22). Naturally, surface roughness is multidimensional and the number of wells is large [~ exp(N)]. With increasing temperature there is a gradual occupation of states, first those with the least energy, then those with higher energy depending on the depths of free energy minima and on the barrier heights. Evidently, the barriers must vary depending on time (on structure, X) and on temperature.* With system cooling in nonequilibrium conditions, the sequence of freezing and state occupancy is inverse. To describe this, Palmer suggested a tree of bifurcations. According to this scheme, below a certain temperature the ground state splits into several (at least in two) states differing by their energy. The process continues to fork as each state appears.

With this scheme it is possible to understand properly the peculiar features of spin glasses in terms of the general conception of glass nonergodicity. This problem is treated by Jäckle,[113] where one can find necessary references. Stillinger,[124] via the scheme presented in Figure 22, preceives the possibility of spatial localization of perfect areas with lower free energy within a prevailing region of higher free energy. This corresponds to local ordering in the amorphous state.

* That must lead to breaking of simple exponential laws with respect to time and temperature relaxation.

Obviously, nonergodicity in the scheme given by Palmer depends on the time scale and temperature. The scheme does not on the whole contradict the thermodynamic concept of the glass transition when a metastable state (i.e., ergodic, $t \to \infty$) gives rise to a frozen-in one (i.e., nonergodic, $t \ll \tau_{rel}$) upon changing the temperature or pressure.

Quantitative calculations for glasses according to the above models that agree with experiment are thus far impossible. The only exceptions are spin glasses. For these, the calculation is possible for the well-known Edwards-Anderson-Ising model which introduced a number of essential simplifications. This model provides quantitative conformity of the calculated statistical entropy as a function of metastable states density, this being defined by their excess energy (Jäckle[174]).

3. Principles of the Description of Glass as a Nonergodic System

According to the correct introduction of statistical probability into a nonergodic system by Jäckle,[174] we shall determine the entropy of glass at $T \to 0$ by the expression

$$S_0(g) = -k_B \sum_i p_i\left(T_{f_i}\right)\ln p_i\left(T_{f_i}\right) > 0 \tag{188}$$

Here i is the configuration index, $p_i\,(T_{f_i})$ is the statistical probability of that configuration at its fictive freezing temperature T_{f_i}. The T_{f_i} temperatures are the same and coincide with T_g in the thermodynamic theory of the glass transition. Equation (188) replaces the Boltzmann formula (187). The entropy and the free energy are functionals, for they depend upon T_{f_i}.

Suppose that for each ith configuration the only state available is that with energy $E_0^{(i)}$. The part of the free energy due to these states at temperature T is expressed as

$$G_T^{(a)}(g) = k_B T \sum_i p_i\left(T_{f_i}\right)\ln p_i\left(T_{f_i}\right) + \sum_i p_i\left(T_{f_i}\right)E_0^{(i)} \tag{189}$$

and at $T \to 0$ as

$$G_0(g) = \sum_i p_i\left(T_{f_i}\right)E_o^{(i)} \tag{190}$$

The entropy at $T > 0$ is greater than that defined by Expression (188) by the amount

$$S^{(b)}(g) = -\partial G_T^{(b)}(g)/\partial T \tag{191}$$

where $G_T^{(b)}(g)$ is the part of glass free energy $[G_T(g) = G_T^{(a)}(g) + G_T^{(b)}(g)]$ associated only with thermal vibrations. This number vanishes at $T \to 0$. The value of $G_T^{(b)}(g)$ is defined by the structure of the glass obtained at T_{f_i} (by means of p_i (T_{f_i}):

$$G_T^{(b)}(g) = \sum_i p_i\left(T_{f_i}\right) \cdot F_i(T) \tag{192}$$

The values of $F_i(T) = -k_B \ln Z_i(T)$ are found using the sum of states for vibrational and other "nonstructural" degrees of freedom, which are predetermined by the given i^{th} configuration frozen in the glass.

Obviously, glasses with different degrees of annealing [i.e., with different sets of $p_i(T_{f_i})$] must have different $G_T^{(b)}$ and different heat capacities, which conforms with experiments. Quantum theories of the heat capacity of glasses thus far do not take this condition into account, since no distinctions with respect to $p_i(T_{f_i})$ or $F_i(T)$ are introduced in these theories.

Lack of ergodicity of a system is widespread and is not a feature only of glasses. However, we shall not concern ourselves here with other aspects of the problem but instead refer the reader to special literature addressing such problems (Palmer,[176] Rostiashvili et al.,[177] and others).

IV Order-Disorder Correlations in the Structure of Glass

A. CALCULATION OF THE EXCESS ENTROPY

1. THE SIMPLEST MODELS

The problem of the structure of glass could become the subject of a special book. However, for our purposes, we confine ourselves to those aspects of the problem which could be directly connected with the calculation of the excess thermodynamic functions of glass, in comparison with crystal. In this respect, first of all those models of disorder should be considered that allow evaluation of the excess entropy of glass.

In the Zachariasen model,[2] it is assumed that elementary structural polyhedra linked by their vertices form a disordered network. The type of network is not defined quantitatively. According to the hypothesis by Lebedev,[183] glass consists of small crystals (named crystallites) that have no phase boundaries.* Neither the sizes of crystallites nor their number are determined. These two limiting models are, obviously, only of speculative interest.

Pauling and Tolman[184] have suggested that the zero-point entropy of glass be calculated by the Boltzmann relation (187). Kelley,[185] by comparing $S_0(g\text{–}c)$ for vitreous silica, ethanol, propanol, and glycerol, concluded that with growing complexity of the molecule, W in Formula (187) increases.

Gutzow[186] considered the following constituents of the excess zero-point entropy of glasses: (a) the entropy of molecular mixing; (b) the disorder entropy defined by the free volume distribution; (c) the configurational entropy of long polymer chains; and (d) the entropy due to mutual disposition and symmetry of molecules. A quantitative comparison was made for item (b). For a total number of sites in the glass network $(N_0 + n_h)$, where N_0 is the number of occupied sites and n_h is the number of free sites, the number of ways of realizing such arrangements is defined as

*Remember Frankenheim's hypothesis![3]

$$W = \frac{(N_0 + n_h)!}{N_0!n_h!} \tag{193}$$

and

$$S_0(b) = -k_B N_0 \ln\left(\frac{N_0}{N_0 + n_h}\right) - k_B n_h \ln\left(\frac{n_h}{N_0 + n_h}\right) \tag{194}$$

Since n_h is small compared to $N_0 \simeq N_A$, it is important to know $N_0/(N_0 + n_n)$, which corresponds approximately to the relation of glass density to crystal density. The calculated $\Delta S_0(b)$ ($J \cdot K^{-1}mol^{-1}$) of 4.4 (SiO_2), 4.3 (GeO_2), 4.5 (BeF_2), 3.3 (Se), and 3.7 (polyethylene) appeared to be close to the experiment, although for glycerol the calculated value is much lower than the experimental one. Such an approach is evidently not comprehensive.

2. Topological Models

The geometrical approach to the formation of a disordered network allows a spectrum of valent angle deviations from their equilibrium values in crystal. For example, even for constant O–Si–O angles within the $SiO_{4/2}$ tetrahedra, one could imagine network disordered due to distortion of the Si–O–Si angles between tetrahedra having a common oxygen. However, the way of space filling by such tetrahedra remains undefined.

Bell and Dean[187] suggested the following topological model for silica glass. Imagine that we are expanding an existing structure by subsequent joining of tetrahedra. The number of ways in which the next tetrahedron may be joined is limited by the way by which the previous one had been joined. If the tetrahedron was connected with the already existing structure by its vertices, the next one might be joined only to one of the previous tetrahedron vertices. But if the previous tetrahedron had been connected with the basic structure only by two vertices, the joining one might occupy either one or two positions with respect to the former. If the previous tetrahedron had been linked with the basic structure by only one vertex, the joining one might be connected with that in no more than four ways. The entropies corresponding to these possibilities would be the maximum of $k_B\ln1$, $k_B\ln2$, and $k_B\ln4$, respectively.

The conditions of network isotropy and of bond saturation require that the number of free vertices during such structure building would at the average, remain constant. For this reason, the relative frequencies of the appearance of triple-, double-, and single-bonded tetrahedra must equal, respectively, p/2, (1 – p), and p/2, where $0 < p < 1$. Then the most probable entropy of the glass network must be

$$S_0 = k_B[(p/2)\ln 1 + (1-p)\ln 2 + (p/2)\ln 4] \tag{195}$$

which corresponds to 5.8 J · $K^{-1}mol^{-1}$ for the maximum possible excess entropy of silica. Bell and Dean proceeded from experiment using a real model to form the structure. They found that instead of one, two, and four ways of tetrahedra connection to triple-, double-, and single-connected previous tetrahedra, 1, 3/2, and 5/2 ways, respectively, were more justified. This gives

$$S_0 = k_B\left[(p/2)\ln 1 + (1-p)\ln(3/2) + (p/2)\ln(5/2)\right] \tag{196}$$

which corresponds to 3.8 J · $K^{-1}mol^{-1}$. This value is close to the experimental value and explains the similarity of the excess entropies of all "tetrahedral networks."

Using this principle of structure modeling, Bell and Dean,[187] as they claimed, obtained good agreement between calculated and experimental radial distribution functions from X-ray and neutron diffraction studies. The distribution of Si–O–Si valent angles in their model has the maximum at 153°, which differs from that of 144° obtained by Mozzi and Warren[65] and of 147° obtained by Narten[188] by using diffraction methods. Although da Silva et al.[189] obtained an angle of 152°, also from diffraction experiments a magic angle spinning NMR experiment by Dupree and Pettifer[190] gave intermediate values of both the distribution curve and the most probable angle. Using a similar methodology, a more recent experiment by Gladden et al.[191] confirmed the picture of the Si–O–Si angle distribution that had been obtained by Mozzi and Warren. Simulations by Thathachari and Tiller[192] where attention was given to steric principles of the formation of Si–O–Si angle distribution, the most probable value for this angle was obtained as 145°.

The above models and experiments on structure demonstrate convincingly that the vagueness of the problem is not yet clarified; we shall return to this issue in the next section. In addition, it is not obvious technological differences in preparing the glass samples do not affect the experimental results.

The above principle of the topological modeling of structures enables one to also calculate the vibrational density of states and some other useful characteristics. One of the most essential consequences of the topological model is the definition of other peculiarities of the linkage of tetrahedra that are, at first sight, specific only for the crystalline state, though the model corresponds to disordered network. In the crystalline modifications of silica, continuous loops are formed by tetrahedra along the lines of Si–O–Si bonds. These represent the shortest path of tetrahedra that return to the silicon atom selected as the origin. Crystalline modifications include 6-membered rings differing by mutual orientation of rings. In the cristobalite structure, the orientation is such that the holes inside rings form channels, whereas in the quartz structure there are no such channels and 8-membered rings are present. The model by Bell and Dean[193] leads to the formation of open systems of channels permeable to some gases, which agrees with the experiment. The structure consists of 5-, 6-, and 7-membered rings. The rings having lower numbers of members are less probable because they cause large mechanic stresses. Gladden[194] designed a similar model containing 6-, 7-, and 8-membered rings.

In the model suggested by Bell and Dean, there are no dangling bonds (i.e., defects). The estimate of their probable contribution to ΔS_0 made by these authors[187] was ~0.4 $J \cdot K^{-1} mol^{-1}$ provided that there were 0.5% dangling bonds. The number is obviously highly overestimated.

Earlier we have mentioned the ample potentialities of topological models for explaining structural variations in the glass transition region. Undoubtedly, in this sense, the structural variations in a glass being annealed, as well as in liquids at various temperatures, are topology changes associated with loop size, shape, and number. These changes are possible only by switching of Si–O–Si bonds. This process is necessary for changing the topological structure when a tetrahedron passes from one loop to another, the size and mutual disposition of loops being varied. That is why structural changes are always limited by elementary activation steps, which are also responsible for viscous flow. A strictly geometrical interpretation of structural disorder leaves unclear the necessity of connecting kinetic parameters of structural relaxation with viscous flow.

When simulating the structure of glassy metals, the ideas of topological distortion of a system were also used (see, for example, Mosseri and Sadoc[195] and Morris[196]). In this case, other views apply that proceed from the possibility of describing a tightly packed crystal structure by Voronoi polyhedra tight packing. The simplest explication of this approach was given by Zallen.[197] The Voronoi polyhedra are formed according to the principle of spatial separation of that part of space around an atom all the points of which are in the utmost vicinity of this atom. The surface of a Voronoi polyhedron presents the one for which all the points are equally remote from both this atom and the neighboring one. Therefore in a tightly packed structure the faces of a Voronoi polyhedron are perpendicular to the lines connecting neighboring atoms. For an fcc lattice, such a polyhedron is a rhombic dodecahedron. In other structures it is different and becomes more complicated when whole groups of atoms (like $SiO_{4/2}$) are located in the sites of a crystalline lattice. The condition of mechanical stability of a system allows the presence of irregular polyhedra. In glass these have a statistical distribution. The polyhedra occupy space in such a way that there appear lines of distortions (disclinations) (see Nelson[198] and Lichatchev and Mikhailin[199]). The capabilities of such a model for describing the silica glass structure are illustrated in the work by Lichatchev and Shudegov.[200]

River and Duffy[201] supposed that the contribution to the entropy of the topological difference between dense atomic liquid and crystal is not greater than $k_B \ln 2$ when only topologically (but not mechanically) stable configurations are taken into account. For networks formed by elementary polyhedra with the internal central atom coordination number equal to z, this value depends upon z according to the relation

$$S_{top} = k_B \left(\frac{1}{2} z - 1 \right) \ln 2 \tag{197}$$

For SiO_2, z = 4 and $S_{top} = k_B \ln 2$, which is obviously higher than those found experimentally. This result is interesting due to the fact that the entropy of disorder must grow with increasing coordination number.

Gupta and Cooper[202] have suggested a general topological principle for the formation of infinite disordered network structures. This principle takes into account the character of sharing of different types of polytopes by vertices (depending on their internal coordination number), thus generalizing the well-known Zachariasen glass formation rules. The criterion is that the average number of degrees of freedom per vertex should not be less than zero.

B. STRUCTURAL REASONS FOR THE EXCESS ENERGY OF GLASS

The simplest models connect the excess energy of glass with freezing of a certain number of two-level systems characterized by a particular excited state energy. The excess energy depends on cooling rate because the number of frozen-in states is also defined by the height of potential barriers (see, e.g., Kob and Schilling[203]).

Gaskell and Tarrant[204] concluded that the structure calculated by the Bell and Dean model[193] must have an excess energy, with respect to cristobalite, of about 46 KJ · mol^{-1}. Relaxation of stresses in this model reduces the energy difference to ~1.7 kJ · mol^{-1}, which is about 4 times less than the experimental value. In their view, the contribution from the internal distortion of $SiO_{4/2}$ tetrahedra to that value is much less than the main contribution, giving the distortion of Si–O–Si valent angles between tetrahedra.

Nemilov[205] presented the excess energy of vitreous silica as the energy of elastic distortion of all the Si–O–Si angles with respect to their value in the structure of α-quartz. The distortions of the angles and distances within a tetrahedron were regarded as infinitesimal since they coincided with the mean deviations due to thermal vibrations.

The excess energy of glass with respect to crystal was defined by Nemilov, in the linear approximation, as

$$\Delta U(g-c) = Nr_0^2 \cdot K_\beta \cdot 10^{-7} \sum_{\delta\beta} \bar{g}_{\delta\beta} (\beta - \beta')^2 \tag{198}$$

where N is Avogadro's number, r_0 is the distance of Si–O, K_β is the force constant of Si–O–Si (β) angle deformation, and $\bar{g}_{\delta\beta}$ is the mean statistical weight of the β angles which fall within a small interval δβ at varying deviations of the angle from its equilibrium value β′ in crystal (Δβ = β – β′). Obviously,

$$\bar{g}_{\delta\beta} = \frac{V(\Delta\beta)\delta(\Delta\beta)}{\int_0^\infty V(\Delta\beta)d\Delta\beta} \tag{199}$$

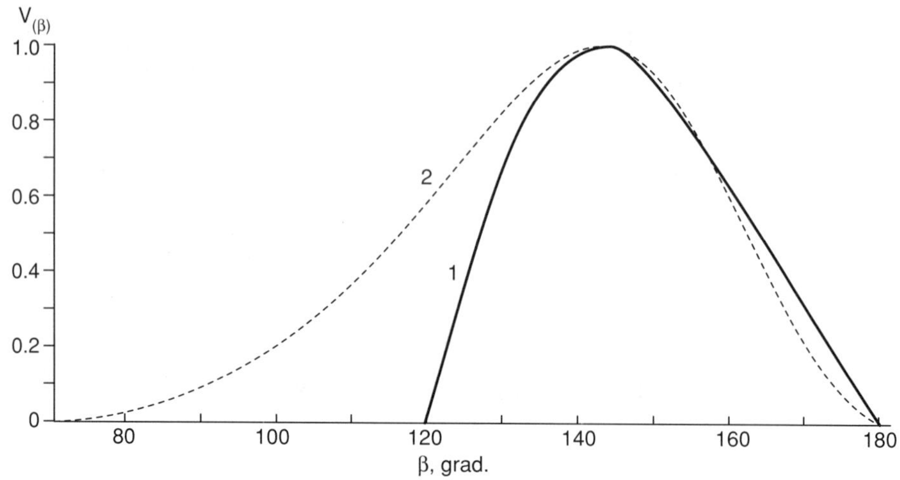

FIGURE 23. Distribution of Si–O–Si angles [V(β)] calculated from the data by Mozzi and Warren,[65] curve 1; and by da Silva et al.,[189] curve 2. (From Nemilov, S. V., *Fiz. i Khim. Stekla (Sov. J. Glass Phys. Chem.),* 8, 385, 1982. With permission.)

where $V(\Delta\beta)$ is the distribution function of the deviations. In Formula (198), $N = 6.02 \cdot 10^{23}$, $K_\beta = 5 \cdot 10^3$ dyn · cm^{-1} (spectroscopy data[206]), $r_o = 1.62 \cdot 10^{-8}$ cm (References 65 and 189) and ΔU is expressed in units of J · mol^{-1}.

Nemilov concluded that in the work by daSilva et al.[189] the glass angle of 152° was determined incorrectly due to attributing the maximum of the Si...Si distance distribution function to this angle. Since $r_{Si-Si} = 2r_0 \sin(\beta/2)$, the distribution function for the angle $V(\beta)$ has its maximum not at 152° but at 144°, as it was estimated by Mozzi and Warren,[65] (see Figure 23). This angle coincides with that of Si–O–Si angle for the α-quartz structure (144°), but differs from that for β-cristobalite (148°). Due to the different shapes of the angle distribution curves in various publications,[65,189] the product of $\bar{g}\delta\beta$ $(\beta-\beta')^2$ yields different integral areas (see Figure 24) and different ΔU values. The value of ΔU(g–c) corresponding to the difference in the energies of glass and α-quartz is, based on the diffraction data by Mozzi and Warren, 4.15 ± 0.3, and, based on the data by da Silva et al., 9.76 ± 0.12 kJ · mol^{-1}. Considering very small thermodynamic corrections, the mean value (6.96 ± 2.93 kJ · mol^{-1}) coincides with that of ΔH_0(g– α-quartz) = 7.19 ± 0.50 kJ · mol^{-1} obtained by the thermodynamic calculation. It should be noted that of all the existing distributions of Si–O–Si angles, those considered in this calculation have the maximum difference in their widths. Apparently, the energy coincidence noted above cannot be accidental.

The estimation by Nemilov[205] of the contribution of structure defects to the ΔH_0 value, which was carried out using the actual concentration and the energy of bond dissociation, yielded a maximum value of about $1 \cdot 10^{-5}\Delta H_0$(g–c); this contribution is small enough to be neglected. Thus we may suppose that at least in such network structures the major contribution to the excess energy of glass comes from the distribution of the valent angles of the bonds that form the network.

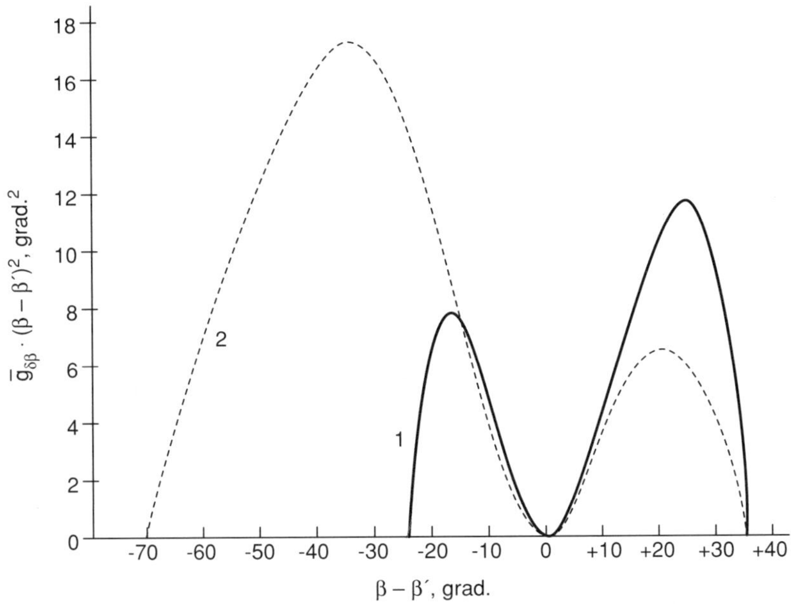

FIGURE 24. $\bar{g}_{\delta\beta}$ $(\beta\text{–}\beta')^2$ as a function in $(\beta\text{–}\beta')$. The curves correspond to the initial data for the V(β) calculations in Figure 23. (From Nemilov, S. V., *Fiz. i Khim. Stekla (Sov. J. Glass Phys. Chem.),* 8, 385, 1982. With permission.)

We have never encountered any excess energy calculations for glasses which, when using model interaction potentials, yielded values comparable with those obtained by thermodynamic experiment. Even using the topological models described above and actual structural data, the resulting values of ΔH_0(g–c) for vitreous silica are about 200 kJ · mol^{-1} and more. Various versions for such calculations are given in the work by Howitt and McElfresh.[207]

C. INHOMOGENEITY OF STRUCTURE AND EXCESS THERMODYNAMIC FUNCTIONS

Here we confine ourselves to one-phase systems, i.e., those where the inhomogeneity is connected with neither crystalline phase transformation nor liquid–liquid phase immiscibility.

Lindenmeyer[208] applied methods of nonlinear nonequilibrium thermodynamics in their general form to interpret the causes for the appearance of glass inhomogeneity. According to Lindenmeyer, glass has excess thermodynamic functions the general form of which, in principle, do not differ from the form of those that have been considered by us earlier. The approach of such a system to equilibrium is described by a dissipative process. If such processes proceed in the same way as dissipative processes in hydrodynamic systems, then density fluctuations of significant amplitude must appear. The excess entropy therefore represents a measure of the nonlinearity or heterogeneity of a thermodynamic system (i.e., glass).

Fujita,[209] applying the principles of statistic thermodynamics, concluded that the free energy minimum is reached by fluctuational ordering within the scale of medium-range order. This phenomenon may result in formation of quasi-crystals, the predecessors of crystal nuclei that have not attained their critical size. In liquid their appearance and disappearance are dynamic, which produces excess entropy due to possible differences in cluster sizes and forms and to the difference in the states of particles in clusters and in the liquid. With decreasing temperature the size of clusters and their fraction in the system would grow. Similar ideas were suggested by Vainas and Almond[210] who estimated the maximum excess entropy for such quasi-inhomogeneous structure as $k_B \ln 2$. Such ideas certainly have a rational basis. In the structural aspect, such an approach was widespread for many decades. However, we do not analyze those works here because of the impossibility of quantitative calculations of the parameters in question.

D. STRUCTURAL INVESTIGATIONS OF THE ORDERING CHARACTER IN GLASSES

The thermodynamic formalism describes glass as a state derived from the state of liquid. With decreasing glass transition temperature, processes of ordering undoubtedly occur in the system. In principle, these may be conceived as striving both to the state of metastable liquid and to the crystalline state. For many decades, researchers using various methods have focused on the determination of the type of these structural processes. To digress from structural processes that are defined directly by the crystalline phase precipitation or by phase separation of the liquid–liquid type, the problem reduces to proving this or that type of ordering in glass as in a one-phase system. Even now it is impossible to give a correct answer to this question.

An interesting discussion took place in Leningrad in 1972 between specialists who were adherents of the crystallite hypothesis of glass structure and scientists who doubted the conclusiveness of the arguments. Considering the capabilities of IR spectroscopy, electron microscopy, and diffraction techniques, the participants in the discussion concluded that none of the methods had a theoretical base for giving convincing information about crystallite's size, shape, and spatial distribution, (Evstropyev and Porai-Koshitz[211]). On the other hand, there is no doubt of the formation in glass of structural units of a certain chemical composition which can have different degrees of geometric ordering. The above applies, obviously, not only to inorganic glasses but also to low-molecular organic glasses and to polymers. For the latter, there were also X-ray structural data available that proved, to the authors' mind, crystallite-like ordering on the scale of 25–30 Å (see, e.g., Yeh[212]).

Simulation calculations do not, on the whole, separate ordering which are defined as short-range or medium-range (Červinka[213]). We have already mentioned some of the difficulties of this approximation.

There have been no recent changes in the situation, in principle. Of significant interest is the low-frequency (i.e., boson) peak of the Raman scattering spectrum (Malinovsky et al.[93,98]). The shape of this peak for melts and glasses is well approximated over a wide frequency region by an exponential correlation function $F(R) \sim \exp(-R/R_o)$, which differs from that of amorphous species of a-Ge type, the latter not being a glass at all. In crystals this peak is absent. According to theoretical models, the F(R) function defines the character of disorder spatial accumulation. At $R \lesssim R_o$, there must exist high spatial correlations of structure. For model systems with regulated variation of the size of crystalline inclusions (L), ω_{max} of that peak changes as $\omega \sim L^{-1}$, which confirms the effect of size. On the other hand, Novikov and Sokolov[214] have found out that the position of the low-frequency (boson) peak for various types of glasses correlates uniformly with the position of the first sharp diffraction peak. This correlation confirms the connection of low-frequency anomalies of vibrational spectra of glasses with characteristic lengths in their structure. Because of insufficient development of the theory, the question of the actual nature of the correlation critical radius R_o and of the unique character of this parameter in the structural aspect remains unresolved.

Wright[215] presented an analysis of the modern state of the theory and experimental capabilities of neutron and X-ray amorphography. In particular, he showed that interpretation of scattering at large and medium angles requires mutual coordination between the details of experiment and the model concepts involved. However, the particular models used (the crystal-based models, random network models, chain and ring models, random sphere packing models, molecular models, amorphous clusters models, Monte Carlo and molecular dynamics simulations, etc.) introduce specific features and never yield comprehensive result. For example, the interference function based on a β-cristobalite model does not agree with the experiment because the Si–O–Si distances and angles are predetermined when passing from short-range order ($r \sim 1.6 + 2.8$ Å) to medium-range order ($r \sim 2.8 + 5.5$ Å) in this model. The Bell and Dean model[193] mentioned previously, besides having similar difficulties, leads to an incorrect distortion of the $SiO_{4/2}$ tetrahedron, and relaxation of the model makes the residual elastic energy too low. However a number of researchers (e.g., Gerber and Himmel[216]) still prefer to use the β-cristobalite structure as the initial one for describing the results of X-ray scattering in SiO_2. That structure is strongly distorted by the presence of non-six-membered rings and topological defects, whose concentration decreases with lowering of the fictive temperature. The most probable Si–O–Si angle appears to be 144°. Apparently, the prehistory of the samples significantly affects the degree of ordering (positron annihilation, Brauer and Boden[217]). X-ray analysis of silica melts and of glasses obtained in different ways lead to the assertion (Nukui et al.[218]) that glass structure may be imagined as if ~ 1/3 of the SiO_2 groups were ordered as in a quartz lattice and the remaining ~ 2/3 as in cristobalite. During the melting process, this relation shifts toward an increase of cristobalite-like ordering.

Investigation of small-angle X-ray scattering enables one to judge the structural inhomogeneities of a sample on the scale of tens and hundreds of angströms. Golubkov[219] has recently shown convincingly that in vitreous SiO_2 there is a structure transition in the region of 1320–1370 K (cooling) and the reverse transition near 1170 K (heating). These states may be related to the formation, at low temperatures, of a quartz-like glass structure modification from a cristobalite-like, high-temperature modification (provided there is one-phase state preservation), since both forms have the feature of a phonon component of small-angle scattering. In connection with this experiment, it would be helpful to remember the work by Arndt and Stoffler.[220] In their case, the investigation of change in properties of vitreous silica at hydrostatic densification (up to 60 kbar) led the authors to the conclusion that a tridimite-cristobalite change of structure occurred near 970 K. The works by Grimsditch[221] and Velde and Couty[222] were devoted to the investigation of the same glass polymorphism caused by greater pressures, but using methods of vibrational spectroscopy. Porai-Koshitz,[223] in reviewing those experiments, mentioned the early work by Tudorovskaya on the stepped change of the refractive index of silicate glasses in similar temperature ranges. He considered the new data on large-angle X-ray scattering in vitreous silica as confirming glass polymorphism.

The information presented here is not comprehensive. Nevertheless, it reveals two circumstances. First, the present methods of describing glass structure based on direct structure investigations cannot yet answer the question about the type of ordering in glass, in the sense of those criteria which we demand (i.e., shape, size, the number of regions of ordering). Obviously, this way of putting the question is too straightforward and primitive, although the structural changes evident from the existence of definite aspects of crystal-like ordering are beyond doubt. Second, another question remains vague, namely distinction between the structure of a glassforming liquid at the melting temperature and that of a crystal at the same temperature when, as it is in SiO_2 case, a glass may have cristobalite-like structure at high temperatures. There is no doubt that the interpretation of changes of thermodynamic functions assuming crystal-like ordering in glass does not contradict well-known structural facts. Further expansion of experimental arguments for this statement is out of place here, for it is fraught with uncritical application of facts.

V Thermodynamic Essence of Glass Structural Relaxation Processes

By the term "glass relaxation" we mean the totality of processes occurring in glass obtained by melt quenching from a certain temperature T_f at the initial equilibrium.* According to Tool,[4] the temperature T_f corresponds to that structural state of a system which had been in an equilibrium liquid but appeared to be frozen in. Obviously, when isothermally sustained, e.g., at $T_1 < T_f$, these processes eventually result in a new equilibrium state of liquid but at the temperature T_1, and the very process shifts T_f from the initial T_f value to $T_f = T_1$. Obviously such a simplified consideration also assumes a unified T_f for all structural processes. However, thus far this limitation is sufficient for solving some basic problems.

In nonequilibrium thermodynamics, for every natural process in systems without mass exchange with their exterior, the entropy production, $\delta_i S$ is always positive,

$$\delta_i S = \delta S - \delta Q / T \tag{200}$$

Here δS is the entropy variation of the system connected with the process and δQ is the heat absorbed by the system from the exterior.[224] This relation is the generalized Second Law of Thermodynamics. For an adiabatically isolated system, $\delta Q = 0$ and the entropy production is equal to the entropy variation of the system; in this case the entropy reaches its maximum. Under isothermal conditions $\delta Q \neq 0$ and

$$\delta_i S = -\delta G_T / T \tag{201}$$

where δG_T is the variation of the free energy (the latter striving to its minimum).

The methods of interpreting an adiabatic process are a bit more complex than that for the case of an isothermal one. However, they can be illustrated explicitly by calorimetric measurements. This is important because they are

* Chemical equilibria displacement is not considered here because that would only introduce an extra parameter but would not change the essential features of the result (Nemilov[129]).

widely practiced. These problems are dwelt on in more detail by Nemilov in Reference 225; here we confine ourselves to a concise survey of the essential issues.

A. ADIABATIC RELAXATION

Suppose a glass having the fictive temperature T_f is instantaneously transferred to an adiabatic shell at the temperature T_1. Two cases are possible, namely, $T_1 < T_f$ and $T_1 > T_f$. The change of the glass state in the structural relaxation process, permitted by the thermodynamic description of such processes,[226] is conventionally divided into two stages: the isothermal enthalpy change at T_1 due to structural relaxation [process (a)] and the subsequent heating or cooling of the relaxation products within the adiabatic shell up to a new temperature T_2 [process (b)].

For estimating the sign of ΔH_{T_1} in process (a), it is necessary to use the full expressions for ΔG_T and ΔS_{T_1} for a chemical reaction using the values of ΔG_0 and ΔS_0 and the respective integral increments for G and S from heat capacities, as described in the previous sections. Considering that the phonon heat capacity below T_1 during the relaxation process varies little in comparison with possible changes of the heat capacity in the T_1–T_f range, we find out that the relaxation process at T_1 is defined for decreasing G_{T_1} and increasing $C_p(g,T_1)$ up to $C_p(msl,T_1)$ by the value of ΔC_p. The sign of ΔH_{T_f} is negative when $T_1 < T_f$ (heat emission) and positive when $T_1 > T_f$ (heat absorption).

In process (b), since the heat capacity of material is always positive and the system is adiabatic, the glass is being heated to the temperature $T_2 < T_f$ (whereas $T_1 < T_f$), i.e., the new temperature $T_2 > T_1$ is established, or cooled to the temperature $T_2 > T_f$ when $T_1 > T_f$ (whereas $T_2 < T_1$). In the calorimetric experiment this would correspond to no work being done because of spontaneous heating for the temperature rise from T_1 to T_2,* the measured heat capacity being zero in the range of T_1–T_2 (the $T_1 < T_f$ case). For the case of $T_1 > T_f$, on the contrary, a surplus amount of work within the shell is necessary because T_2 is lowered as compared with T_1 due to the relaxation process. In this case, the measured heat capacity appears overstated. The above correlations are illustrated by the scheme in Figure 25. Under real conditions, because of the continuity of heating and the spectrum of relaxation times, these thermal effects are smeared out and the heat capacity curve shows a little "fall" before the heat capacity increases, and then an "urspurge" of the heat capacity on the peak of its jump (see Figure 26). Obviously, the magnitudes of these effects are connected with the dynamics of the temperature rise and with the extent of the annealing of the glass at T_1 below the initial T_f. As annealing develops, T_f decreases, and at the same rate of the temperature rise the minimum becomes

* An adiabatic shell allows work to be done but no heat influx, therefore for rising temperature it is used to calculate, e.g., the work done by electric current. The DSC method is somewhat different from the adiabatic one.

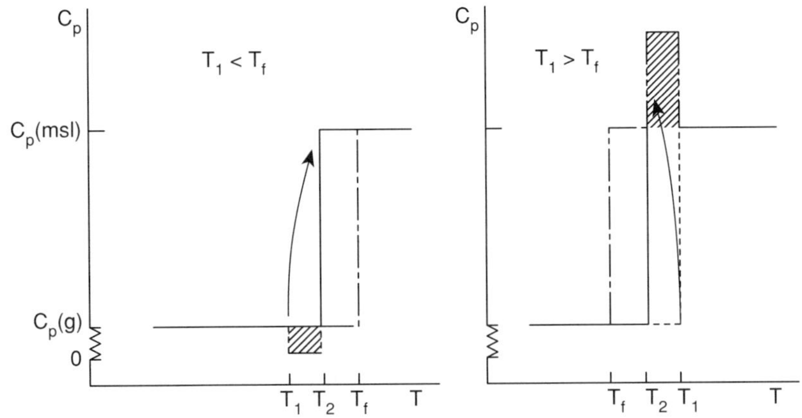

FIGURE 25. Effective heat capacity formation for adiabatic relaxation of glass under conditions of $T_1 < T_f$ and $T_1 > T_f$. The solid line corresponds to the idealized effective heat capacity due to exo- and endo-effects. Arrows indicate the evolution of the real heat capacity and the temperature due to structural relaxation. The shaded area corresponds to the magnitudes of exo- and endo-effects. (From Nemilov, S. V., *Fiz. i Khim. Stekla (Sov. J. Glass Phys. Chem.),* 11, 146, 1985. With permission.)

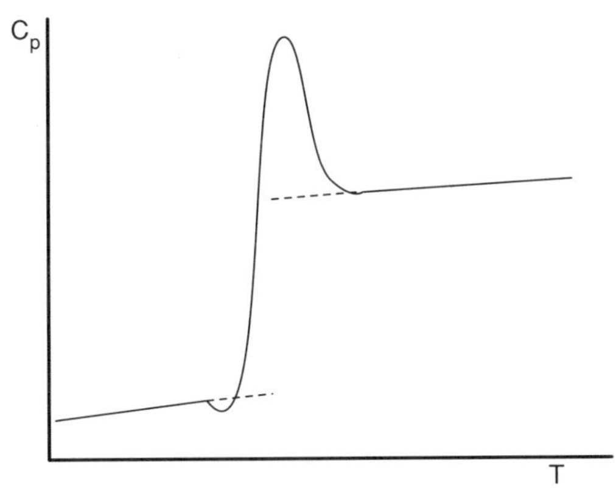

FIGURE 26. Real effective heat capacity measured by an adiabatic calorimeter.

deeper and the maximum higher. Both extrema may shift toward higher temperatures because the viscosity increases due to the annealing process. As the heating rate increases, at a constant initial fictive temperature, both minimum and maximum are shifting to the right whereas the minimum deepens and the maximum grows. All of these phenomena have been thoroughly studied (see References 130, 134, 135, 227, 228).

The above phenomena possess at the same time both a relaxational and a thermodynamic nature and are specific only for nonequilibrium systems. The heat capacity measured in the regime of decreasing temperature never exhibits such features, and its curve has the shape of a smooth step. This shape is defined by the relation of the widths of the structural relaxation time distribution to the cooling rate. That is why the heat capacity upsurges obtained in adiabatic experiments are side effects and are not to be taken into account in calculations of the absolute magnitude of the heat capacity jump ΔC_p at the glass transition.

The heat capacity anomalies are the simplest examples of the "memory effect" of glass. Glass memorizes the thermal history that has defined its internal structure and, hence, the values of the free energy and the entropy at the absolute zero. This memory is manifested in a qualitative form affecting the heat capacity only when the kinetic and temperature conditions become appropriate for the system's response. These conditions (in particular, the temperature) are close to those at which the system has been memorizing information.

Effects like the above are observed for volume relaxation (i.e., the thermal expansion coefficient) and other properties of glass which are connected with its structure (see, e.g., Mazurin[229]).

B. ISOTHERMAL RELAXATION

The case of isothermal relaxation is defined by the relation (201) and is simpler than the previous one. As was shown by Nemilov,[225] entropy production at $T = T_1$ is defined by two terms:

$$\delta_i S = \delta_i S(\text{entr}) + \delta_i S(\text{heat}) \tag{202}$$

$$\delta_i S(\text{entr}) = -\int_{T_1}^{T_m} \delta C_p \, d\ln T \tag{203}$$

where $\delta_i S(\text{entr})$ defines the entropy change at the structural relaxation (this value is also connected with the variation of the zero-point entropy), and

$$\delta_i S(\text{heat}) = -\frac{1}{T_1}\int_{T_1}^{T_m} C_p(g)\, dT \tag{204}$$

where $\delta_i S$ (heat) defines the heat content change at the structural relaxation (this value is also connected with the variation of the zero-point free energy). The integral in Equation (204) is determined from independent experiments as the difference between the crystallization heats at T_1 for quenched and annealed glasses. During the annealing process, $\delta C_p \to \Delta C_p$ and the signs of $\delta_i S(\text{entr})$ and $\delta_i S(\text{heat})$ change depending on the relation, either $T_1 < T_f$ or $T_1 > T_f$ (i.e., exo- or endothermal effects).

Entropy production, according to Equation (201), is always positive because the free energy of glass is always more than that of metastable liquid.

For deviations of T_1 from T_f defined by the inequalities

$$0.8 \le T_1 / T_f \le 1.2 \tag{205}$$

and for deviations of no more than 10%, $\delta_i S$ may be calculated approximately by the relation

$$\delta_i S = \frac{\Delta C_p (T_f - T_1)}{T_1} f\left(\frac{T_1}{T_f}\right) \tag{206}$$

Here, depending on the mathematical approximation assumed for the expansion of $\ln(T_f/T_1)$, $f(T_1/T_f)$ may have the form of

$$f_1(T_1 / T_f) = 1 - (T_1 / T_f) \tag{207}$$

and

$$f_2(T_1 / T_f) = (T_f - T_1)/(T_f + T_1) \tag{208}$$

ΔC_p is presumed to be constant in the T_1–T_f region. Using those approximations, one can evaluate δG_{T_1}, the driving force of the relaxation process, from Equation (201).

C. STRUCTURAL RELAXATION AND DENSITY FLUCTUATIONS

By its outward appearance, the experimental course of the heat capacity being measured in an adiabatic calorimeter above T_g looks like a phase transition of the second order. The resemblance is even more enhanced by the fact that the value of the upsurge, depending on the measurement regime, may become rather significant. Thermodynamic considerations show that this phenomenon is due to the system's loss of thermodynamic stability. The latter, however, is only associated with kinetic limitations. The system passed from a frozen state into an equilibrium one but by no means from one equilibrium state into another, as would have occurred at a phase transition.

As for any unstable state, the state of glass must be characterized by the violation of the inequalities $dT/dS > 0$ and $dp/dV < 0$. These mean that each pair of sign-defined changes in the thermodynamic variables (dT and dp) leads either to an entropy decrease (instead of its growth for an equilibrium system, in our case at heating below T_f) or to a volume increase (instead of its reduction, in our case at heating above T_f).

Above T_f, irreversible local changes of structure and the transition of the entire system into a metastable state by the evolution of fluctuations are possible. This corresponds to the increase of density fluctuations having long-range correlations, i.e., to abnormally large transient compression effects. These, according to fluctuation theory, as for the case of the second-order phase transitions,[230] must be followed by abnormally large density fluctuations and light-scattering anomalies. These must gradually vanish with time as the

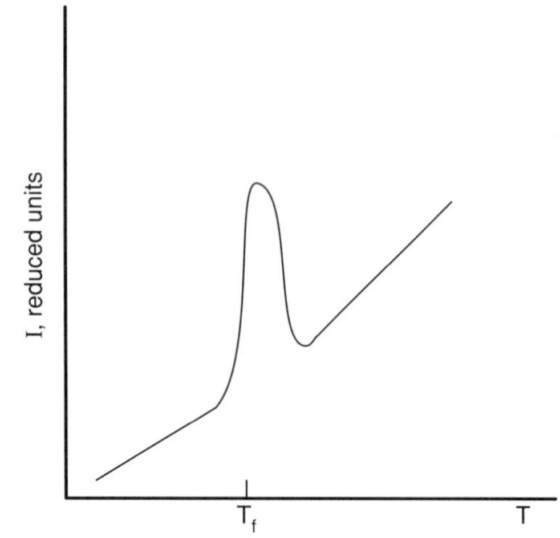

FIGURE 27. Expected change of the intensity of electromagnetic waves scattering in the vicinity of T_f for a glass heated at a certain rate. (From Nemilov, S. V., *Fiz. i Khim. Stekla (Sov. J. Glass Phys. Chem.),* 11, 146, 1985. With permission.)

equilibrium is approached. This consideration is similar to Lindenmeyer's view of the appearance of inhomogeneous glass structure,[208] but used here to interpret the present phenomena.* It caused Nemilov[225] to suppose that the electromagnetic waves scattering in the T_f vicinity have an upsurge as shown in Figure 27. The amplitude of the upsurge must reduce to zero with time and depend on the heating rate and the sample prehistory, as well as the heat capacity upsurge.

This very effect had been discovered by the method of small-angle scattering by Vasilevskaya et al.[231] for B_2O_3 and was studied in detail using light-scattering methods by Bokov[232] for glasses in the borate, silicate, and germanate systems. As for the heat capacity, the upsurge amplitude increases with the extent of annealing. It was established that annealing results in the formation of doublet and triplet splittings and to the polarization of the scattering components. Although, on the whole, these peculiarities of the relaxational scattering upsurge remain obscure, the very fact of the polarization is evidence, to our mind, of the tensor nature of the structural parameter of ordering.

D. THE MEMORY EFFECT IN CROSSOVER EXPERIMENTS

The condition for the existence of a spectrum of structural relaxation times in the above calorimetric and dilatometric experiments manifests itself in a rather hypothetical form as a condition of endo- and exo-effects being "blurred,"

* *Addition in Proof:* These fluctuations are transient but their origin may be of the same nature as the appearance of large stable fluctuations due (to) bifurcations, in new Prigogine's non-linear irreversible thermodynamics.

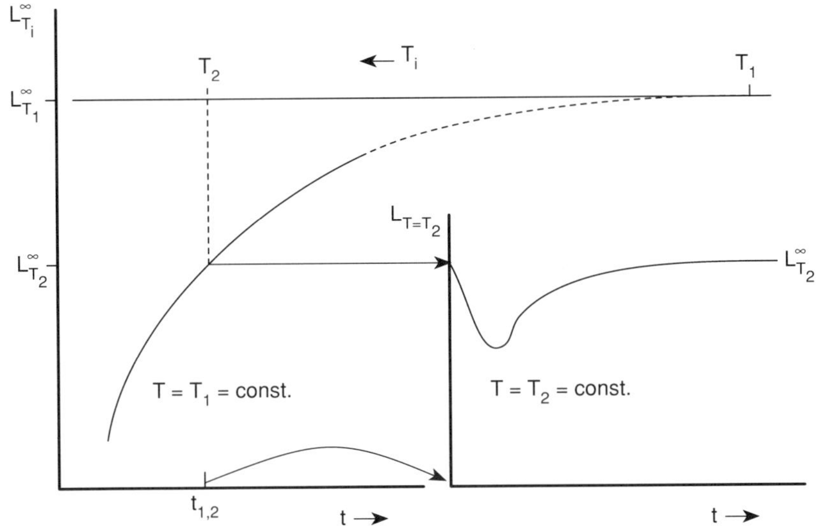

FIGURE 28. "Memory effect" of glasses in the "crossover" experiments (for an explanation, see the text). (From Nemilov, S. V., *Fiz. i Khim. Stekla (Sov. J. Glass Phys. Chem.),* 12, 293, 1986. With permission.)

whereas there are a number of experiments for which a sufficiently wide spectrum of relaxation times is necessary to explain the results.

Experimental data relative to this problem are noteworthy.[233–238] When in the stabilization process of a quenched glass a property L* at temperature T_1 varies smoothly as it approaches the equilibrium value $L^{\infty}_{T_1}$, it runs through a number of values L_{Ti}. Each of these values corresponds to a certain equilibrium value of $L^{\infty}_{T_i}$ at a certain temperature T_i, where $T_i > T_1$. If the annealing process has been interrupted at that exact value of the property (or at a nearby value) and the system has been rapidly brought to the temperature $T_2 = T_i$, the property does not appear to be constant. The L_{T2} value continues to change with time, and only after passing through its minimum does it begin to increase approaching the equilibrium value $L^{\infty}_{T_2}$ (see Figure 28). The direction of the temperature jump, which defines the direction of the property change may vary. It only modifies the relaxation kinetics character when such a process has been interrupted and continued at an intermediate temperature.[239] In any case, the system memorizes the thermal history that has preceded the last stage of thermal treatment. The chemical nature of the vitreous state is apparently of no importance for the manifestation of the effect. The kinetics of, e.g., volume relaxation, which defines the refractive index, is connected with the superposition of rapid and slow process. Florinskaya[233] came to such a conclusion based on experiments carried out in the 1940s on lead flints and barite crowns.

* This may be any structure-sensitive property. Usually, either the volume or the associated refractive index is in question. Refractive index measurements can be carried out with the utmost precision.

To explain crossover experiments, it is necessary to presume[233,236,239] that the value of the property L is defined by at least two parameters, ξ_1 and ξ_2. These are able to relax at different rates, so different fictive temperatures, $T_{f_1}(\xi_1)$ and $T_{f_2}(\xi_2)$ may be attributed to them. Both parameters make approximately equal (comparable) contributions into the property value, and the time dependence of the property is defined by the superposition of the T_{f_1} and T_{f_2} time dependences,

$$\frac{dL}{dt} = \frac{\partial L}{\partial T_{f_1}} \cdot \frac{\partial T_{f_1}}{\partial t} + \frac{\partial L}{\partial T_{f_2}} \cdot \frac{\partial T_{f_2}}{\partial t} \tag{209}$$

In the initial state of stabilized glass (T_0), dL/dt = 0 since both relaxation processes have been completed. In the final state, (at the temperature T_1, see Figure 28) dL/dt = 0 also. At $t_{1,2}$ (before complete equilibrium has been reached), because of the higher rate of one of the processes (say, the one connected with a T_{f_1} shift), this process may have been completed ($\partial T_{f_1}/\partial t \simeq 0$). At that time, the process connected with a T_{f_2} shift would not yet have been completed. For a temperature intermediate between the initial T_0 and T_1 temperatures, the rapid process associated with a T_{f_1} shift would proceed in a direction opposite to that of the slow one (as $|T_1 - T_0|$ is less than $|T_2 - T_0|$), but in that very direction, as before. The curve L = L(t) shows a minimum after passing the time $t_{1,2}$. The minimum corresponds to the equality of the absolute values of the terms in the right-hand side of Relation (209), which have different signs. Macedo and Napolitano[236] assumed that the mean time of volume relaxation in such experiments for a borosilicate crown was defined by the following superposition of rapid and slow process relaxation times,

$$\bar{\tau}_{pT} = \sum g_i \tau_i = (\tau_1)/2 + (\tau_2)/2 \tag{210}$$

while a crossover experiment is well described at $\tau_2/\tau_1 = 8$. For other glasses (GeO_2,[240] B_2O_3[237]), the ratio τ_2/τ_1 is of the same order. In every case, the activation energies defining $\bar{\tau}_{pT}$ are very close to the effective activation energies for viscous flow. Comparison of the relaxation time $\bar{\tau}_{pT}$, the relaxation time for purely shear relaxation, and the adiabatic isochoric relaxation time, enabled the authors[241] to state that the expected relation of the relaxation times for shear and volume viscosities for associated liquids is retained, and all the observed relaxation times interrelate with mechanic relaxation. Thus, both times (τ_1 and τ_2) may be referred to the same process, which is characterized by a relaxation time distribution. The width of this distribution reduces with the temperature increase, as was shown for a shear viscosity relaxation time distribution (e.g., B_2O_3, Macedo and Litovitz[241] and Boesch et al.[237]). Now as we shall see later, it is known that, for similar mean relaxation times of different processes defined by the relaxation of different structural parameters (the parameters of ordering), there is a difference in the exact characteristics of the shape and the width of the relaxation time distribution. Thus far there is apparently no absolutely reliable guarantee that the τ_1 and τ_2 times cannot be referred to the relaxation of different but still mutually dependent order parameters.

The microscopic interpretation suggested by Macedo and Litovitz,[241] based on the chemical inhomogeneity of borosilicate glass can hardly be comprehensive. Such memory effects are observed in B_2O_3 and GeO_2 glasses, which have no chemical inhomogeneity, and in polymers as well.[242] Cooper[243] considered a topological model of vitreous B_2O_3. He presumed that the difference in the kinetics may be due to the difference in the relaxing parameters, whose variation is defined by the process of switching of B–O–B bonds (this must be of greater relaxation time) and by changes in the angular relations of the bonds. Nemilov,[239] for the same purpose, invoked the model of micro- and macroviscosity, which is used in the molecular spectroscopy of solutions to explain fluorescence kinetics. Applying it to glass, this model reduces to Cooper's one.

E. THE PRIGOGINE-DEFAY RATIO AND THE RELAXATION KINETICS OF THERMODYNAMIC PROPERTIES

1. Formal Aspects of the Description of Relaxation Kinetics

The description of relaxation kinetics in terms of fictive temperature drifting under conditions of the possible existence of several structural parameters needs explanation. If a property L has the value L^0 at the beginning of the process (t = 0) and the value L^∞ at the end ($t \to \infty$), the current value of the property L^t is related to the L^0 and L^∞ values by the relaxation function

$$\frac{L^t - L^\infty}{L^0 - L^\infty} = \phi(t) \tag{211}$$

In the beginning of the process $\phi(0) = 1$ whereas on its completion $\phi(\infty) = 0$. Presumably, the variation of a thermodynamic parameter that causes the process (e.g., temperature or pressure change, etc.) is small enough to preserve the linear approximation, and during relaxation the exterior thermodynamic parameters (e.g., the temperature and the pressure) remain constant. For one relaxation time τ_0, $\phi(t)$ corresponds to a simple exponential of the form

$$\phi(t) = \exp(-t/\tau_0) \tag{212}$$

If several (N) parallel relaxation processes are possible, each having the relaxation time τ_i, then

$$\phi(t) = \sum_{i=1}^{N} g_i \exp(-t/\tau_i) \tag{213}$$

where g_i is the statistical weight of the ith process and $\sum_{i=1}^{N} g_i = 1$. If the spectrum of the relaxation times is continuous, then

$$\phi(t) = \exp\left[-\left(t/\tau_K\right)^{\beta}\right] \tag{214}$$

where τ_K is not the relaxation time of a specific process because the very dependence is fractal, the fractal coefficient β being defined by the inequality $0 < \beta \leq 1$.* The form of the dependence in (214) may, in principle, be derived mathematically from (213) (see Reference 244). In the approximation of small deviations from equilibrium, the relaxation function is an equilibrium one. The average relaxation time is then connected with τ_K by a gamma-function,

$$\langle\tau\rangle = \tau_K \Gamma(1/\beta)/\beta \tag{215}$$

The relaxation time spectrum corresponding to (214) is asymmetric. It should be noted that a dependence like (214) may be derived as well for a sequence of correlated activation steps (see, e.g., Palmer et al.[245]). Equation (214) was suggested as an empirical one by Kohlrausch[246] as long ago as 1854 to describe the decay of charge in Leiden jars and later was used by him and his son for the description of mechanical relaxation in glasses and other phenomena. After a long period of oblivion, it was used by Williams and Watts[247] for the description of dielectric relaxation in polymer melts near T_g. Their work cleared the way for widespread use of that relation. Relation (214) provides a popular description of relaxational phenomena in glasses near T_g (α-relaxation). Now it is practically impossible to find relaxational experiments that under specific conditions, cannot have be approximated by this relation with enough precision. The values of τ_K and β depend on temperature, the τ_K values varying approximately like viscosity whereas β approaches 1 with increasing temperature (which means spectrum narrowing).

Equation (211) is also used in studying enthalpy relaxation. If the enthalpy has a current value $H^t(T)$, to which fictive temperature T_f relaxation (drifting) corresponds, then

$$H^t(T) = H(msl)\left(T_f\right) - \int_{T}^{T_f} C_p(g)dT' \tag{216}$$

where $H(msl)(T_f)$ is the equilibrium enthalpy at T_f (it corresponds to metastable liquid).** The temperature dependence of T_f is defined by heat capacity relaxation,

* Dependence (214) is also named "the stretched exponential law."

** In Reference 227 the reader will find further references concerning the basis of this formalism. These include those by R. O. Davies, G. O. Jones, G. Gardon, O. S. Narayanaswamy, C. T. Moynihan, H. N. Ritland, and others.

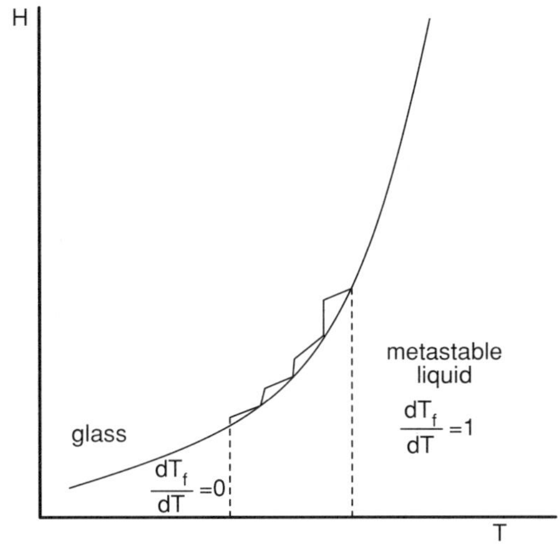

FIGURE 29. Illustration of Relations (216) and (217) for cooling by four temperature jumps. For continuous cooling their number becomes infinite and Relations (216) and (217) begin to hold.

$$\frac{dT_f}{dT} = \frac{\left[C_p - C_p(g)\right]_T}{\left[C_p(msl) - C_p(g)\right]_{T_f}} \tag{217}$$

These relations mean that the heat capacity at the glass transition exhibits a smooth dependence. The integral in (216) is the sum of stepped changes in the heat content provided the temperature change is presented by jumps. At each jump $T_f \rightarrow T$ the heat content decreases (Figure 29). The time dependence of T_f is bound up with the relaxation function by the relation

$$T_f(t) = T_f(0) + \delta T\left[1 - \phi(t)\right] \tag{218}$$

where δT is the value of the temperature jump that causes the enthalpy relaxation. If τ_i values depend on δT and on time, an integral relation (213) is needed (Gardon and Narayanaswamy[248]),

$$\phi(t) = \sum_{i=1}^{N} g_i \exp\left(-\int_0^t dt'/\tau_i\right) \tag{219}$$

The temperature dependence of τ_i for a fairly narrow temperature range may be approximated by a distorted Arrhenius exponent,

$$\tau_i = A_i \exp\left[x\Delta H^{\ddagger}/RT + (1-x)\Delta H^{\ddagger}/RT_f\right] \tag{220}$$

where in the first approximation $\Delta H^{\ddagger}$ may be regarded as corresponding to the activation enthalpy of viscous flow and x is defined by the relation of the viscous flow activation enthalpy for the glass with fictive temperature T_f to that of the equilibrium liquid. Relation (220) presumes all the relaxational processes of the ith type have the same activation energy, therefore

$$\tau_K = A \cdot \exp\left[x\Delta H^{\ddagger} / RT + (1-x)\Delta H^{\ddagger} / RT_f\right] \tag{221}$$

For τ_K, Relation (221) produces the same dependence given by the Arrhenius law at $T_f = T$. DeBolt et al.[227] derive other special equations by which calorimetry can be used for studying structural relaxation in a dynamic temperature regime. They also derive a relation for the T_f dependence on the rate of melt cooling,

$$d\ln|q| / d(1/T_f) \simeq \Delta H^{\ddagger} / R \tag{222}$$

Analogous relations apply for relaxation of volume and other properties.

It should be mentioned that when studying volume relaxation in polymers another formalism, which was suggested by Kovacs (see e.g., Reference 249), is often used. In this case the deviation from equilibrium is not Function (211) but the relation

$$\delta = \left(v - v^{\infty}\right) / v^{\infty} \tag{223}$$

where v^{∞} is the volume under equilibrium conditions and v is the current volume value. The rate of relaxation is defined by the relation

$$-d\delta / dt = q\Delta\alpha + \delta / \tau \tag{224}$$

where q is the cooling rate, $\Delta\alpha$ is the change of the thermal expansion coefficient at glass transition, and τ is the relaxation time. For the each ith relaxational process

$$-d\delta_i / dt = \Delta\alpha_i \cdot q + \delta_i / \tau_i \tag{225}$$

where δ_i, $\Delta\delta_i$, and T_i have the same meaning but refer to the ith process. Their statistical weights are normalized to 1. The dependence of τ_i vs. T and δ_i is as follows:

$$\tau_i(T,\delta) = \tau_{i,r} \exp\left[-\theta / \left(T - T_r\right)\right] \exp\left[-(1-x)\theta\delta / \Delta\alpha\right] \tag{226}$$

where $T_{i,r}$ is the relaxation time of the ith process at the reference temperature T_r and θ and x are material parameters. These define the temperature dependence of the relaxation time and the relative contributions of the temperature and structural changes to τ_i.*

* In the Narayanaswamy and Moynihan theory, the analogous parameters are $\Delta H^{\ddagger} = \theta RT^2$ and x, respectively.

This approach also involves the fictive temperature parameter; according to Tool,

$$T_f = T + \delta / \Delta\alpha \tag{227}$$

whereas

$$dT_f / dt = \left(T - T_f\right) / \tau \tag{228}$$

Enthalpy relaxation has a similar behavior. Since the theory presumes a multiparameter approximation, it successfully describes memory effects in polymers, which we have discussed above, and the dependence of T_g on cooling rate. It also explains the appearance of the stretched exponential law. The temperature dependence of τ_i may be expressed in terms of any adequate relaxation theory. We do not discuss this theory in more detail here, since the basic experiment, which is discussed below, has been treated by applying the Narayanaswamy and Moynihan formalism. The result would apparently be the same as that of the Kovacs formalism.

2. The Relation Between Relaxation Functions and the Prigogine-Defay Ratio

In describing changes in glass properties with time, it is convenient to apply the notion of a fictive temperature T_f. It is especially helpful in engineering calculations. Essentially the fictive temperature is a certain generalized structural parameter whose deviation from the value of the thermodynamic temperature defines the rate of relaxation of that parameter [Tool,[4] Equation (228)]. As we have seen, it enables one to take into account easily the nonlinearity of the relaxation due to the dependences of the relaxation time on temperature and on the prehistory of the material. The single-parameter approximation does not adequately correspond to reality because the value of the Prigogine-Defay Ratio is practically always more than 1. Several fictive temperatures may be included, each of which should be correlated with some particular property. It would also be natural to introduce another characteristic of structure, the fictive pressure p_f. Such an approach was applied by Gupta.[250] For the general consideration of the thermodynamic aspects of relaxation, it would be still more conveniently studied in terms of the formalism involving many dimensionless structural parameters. In principle, such an approach for the particular group of closely related property parameters would lead to the formalism of fictive temperature or fictive pressure. This is clear from the fact that for a single-order parameter the relaxation law is similar to that for fictive temperature relaxation (228). The main difficulty is in specifying the physical features of each structural parameter in the sense of its effect on each macroscopic property.

Earlier, when determining the properties of the order parameter ξ, it was postulated that the relation specifying the linear law of relaxation rate vs.

thermodynamic affinity ($-G_\xi$), Equation (44), was

$$d\xi / dt = -\gamma G_\xi \tag{229}$$

where γ is a kinetic coefficient. The exponential law of relaxation, Equation (46),

$$d\xi / dt = -(\xi - \xi^e) / \tau \tag{230}$$

in which a current ξ value is related to the equilibrium one (ξ^e) by the relaxation time, τ, is equivalent to this correlation. Relation (47),

$$\tau = (\gamma G_{\xi\xi})^{-1} \tag{231}$$

also holds. Consider relaxation caused by a small, sudden variation of temperature, δT, or pressure, δp. The relaxational function for the free energy relaxation is expressed as

$$\frac{G^t - G^\infty}{G^0 - G^\infty} = \phi(t) \tag{232}$$

We designate $G^t - G^\infty = \delta G^t$, $G^0 - G^\infty = \delta G^0$, $\xi^0 - \xi^e = \delta\xi^e$, $\xi^t - \xi^e = \delta\xi^t$. The t and 0 superscripts correspond, respectively, to the current values at the moment t and to the initial values at t = 0. We recall the designation meanings: $G_\xi = \partial G/\partial\xi$; $G_{\xi\xi} = \partial^2 G/\partial\xi^2$; $G_{\xi T} = \partial^2 G/\partial\xi\partial T$; $G_{\xi p} = \partial^2 G/\partial\xi\partial p$. The relation between $\delta\xi$, δT, and δp variations in the neighborhood of equilibrium was defined by Equation (58),

$$\delta\xi = -\left[(G_{\xi p} / G_{\xi\xi})\delta p + (G_{\xi T} / G_{\xi\xi})\delta T\right] \tag{233}$$

Using the latter, one can easily find the correlation between these variations for the cases of pressure or temperature constancy.

Obviously, the numerator in the left-hand side of Equation (232) may be defined as

$$\delta G^t = \delta(G_\xi)_p \cdot \delta\xi^t \quad (p = \text{constant}) \tag{234}$$

or

$$\delta G^t = \delta(G_\xi)_T \delta\xi^t \quad (T = \text{constant}) \tag{235}$$

The denominator in Equation (232) may be defined as

$$\delta G^0 = \delta(G_T)_p \cdot \delta T \quad (p = \text{constant}) \tag{236}$$

or

$$\delta G^0 = \delta\left(G_p\right)_T \cdot \delta p \quad (T = \text{constant}) \tag{237}$$

where $\delta(G_T)_p$ and $\delta(G_p)_T$ are the values of affinity produced by drastic temperature or pressure change (per unit change of the temperature or pressure):

$$\delta\left(G_T\right)_p \equiv -\left(G_\xi\right)_p \left(\partial\xi / \partial T\right)_p \tag{238}$$

$$\delta\left(G_p\right)_T \equiv -\left(G_\xi\right)_T \left(\partial\xi / \partial p\right)_T \tag{239}$$

The relaxation function (232) is expressed as

$$\phi(t) = \frac{G_\xi \cdot \delta\xi^t}{\delta\left(G_T\right)_p \cdot \delta T} \quad (p = \text{constant}) \tag{240}$$

Substituting the relation between δT and $\delta\xi^0$ from (233) into (240), we have

$$\phi(t) = \left[-\frac{G_\xi \cdot G_{\xi T}}{\delta\left(G_T\right)_p G_{\xi\xi}}\right] \frac{\delta\xi^t}{\delta\xi^0} \tag{241}$$

Here $\delta\xi^0$ corresponds to $\delta\xi^t$ at $t = 0$.

Obviously, by virtue of Relation (230),

$$\partial\xi^t / \delta\xi^0 = \exp(-t/\tau) \tag{242}$$

and the factor [ ] in the right-hand side of Relation (241) is equal to 1 by virtue of Relation (238). A similar conclusion is correct for the case of T = constant.

These considerations clarify the reasons why factors like $(G_\xi)_p$ $(\partial\xi/\partial T)_p$, etc. may appear in the expression for the relaxation function. In the general case, in the analysis of the relaxation of a thermodynamic property Q caused by the variation δX of a thermodynamic variable X, the expression in the denominator of the relaxational function (232) is $-Q_\xi(\partial\xi/\partial X) \cdot \delta X^0$ and the expression in the numerator of that function is $Q_\xi \cdot \delta\xi^t$.

Relaxation associated with the free energy relaxation with T and p variables may be investigated in four experiments: (a) entropy (enthalpy) relaxation stimulated by the temperature variation δT (constant pressure); (b) volume relaxation stimulated by the pressure variation δp (constant temperature); (c) volume relaxation stimulated by the temperature variation δT (constant pressure); and (d) entropy (enthalpy) relaxation stimulated by the pressure variation δp (constant temperature).

For these experiments, factors appear in the expressions for the relaxational function that define the increments of the thermodynamic coefficients (i.e., of the heat capacity ΔC_p, compressibility $\Delta\kappa_T$, and thermal expansion $\Delta\alpha$) at the glass transition:

$$\text{(a)}\quad \frac{\partial S}{\partial \xi}\cdot\frac{\partial \xi}{\partial T} = -G_{\xi T}\left(\partial\xi/\partial T\right)_p = \frac{\Delta C_p}{T} \tag{243}$$

$$\text{(b)}\quad \frac{\partial V}{\partial \xi}\cdot\frac{\partial \xi}{\partial p} = G_{\xi p}\left(\partial\xi/\partial p\right)_T = -V\Delta\kappa_T \tag{244}$$

$$\text{(c)}\quad \frac{\partial V}{\partial \xi}\cdot\frac{\partial \xi}{\partial T} = G_{\xi p}\left(\partial\xi/\partial T\right)_p = V\Delta\alpha \tag{245}$$

$$\text{(d)}\quad \frac{\partial S}{\partial \xi}\cdot\frac{\partial \xi}{\partial p} = -G_{\xi T}\left(\partial\xi/\partial p\right)_T = -V\Delta\alpha \tag{246}$$

The meaning of these is clear from the first derivatives of the thermodynamic property in the parameter ξ being defined by the second derivatives of the free energy, i.e., $|Q_\xi| = |G_{\xi X}|$.

In the general case, simultaneous relaxation of many parameters of order ξ_i is possible, each relaxing with its own relaxation time τ_i. It is presumed that all of the $G_{\xi_i\xi_i}$ and the kinetic coefficients related by Equation (231) are nonzero, and coefficients such as $G_{\xi_i\xi_j}$ are zero ($G_{\xi_i\xi_j} = 0$, at $i \neq j$). In this case the relaxation function may be expressed as

$$\phi_{QX} = \sum_i g^i_{QX}\exp\left(-t/\tau_i\right) \tag{247}$$

where the weighting coefficients are normalized such that

$$\sum_i g^i_{QX} = 1 \tag{248}$$

For each parameter, the weighting coefficient is defined as

$$g^i_{QX} = \frac{\left(G_{\xi_i X}\right)^2}{\sum_i G_{\xi_i X}\left(\frac{\partial \xi_i}{\partial X}\right)G_{\xi_i\xi_i}} \tag{249}$$

For this general case we cannot do cancellations such as we did in Expression (241) for single-parameter relaxation. In the denominator of the relation is the sum of products of coefficients that form a positively defined matrix. Relations (243)–(246) lead to the Prigogine-Defay Ratio and determine the criterion of the stability of the final completely relaxed state. For weighting coefficients defining different property relaxation, the relation

$$\frac{g^{i}_{VT}\, g^{i}_{Sp}}{g^{i}_{Vp}\, g^{i}_{ST}} = \Pi \tag{250}$$

holds for the same parameter ξ_i. For the case of linear mutual correlation of all order parameters or the sole parameter, when $\Pi = 1$ [see Equation (72)], $\phi_{VT} = \phi_{Vp}$, $\phi_{VT} = \phi_{ST}$, etc. hold.

If $\Pi > 1$ and there is more than one independent order parameter, $\phi_{VT} \neq \phi_{VP}$; $\phi_{VT} \neq \phi_{ST}$, etc., but $\phi_{VT} = \phi_{Sp}$. For this case, using the Cauchi-Schwarz inequality at the introduction of auxiliary functions (see References 31, and 251 to 253), it appears that the relaxational function relation defined as

$$M \equiv \frac{\phi_{VT} \cdot \phi_{Sp}}{\phi_{Vp} \cdot \phi_{ST}} \tag{251}$$

possesses the following property:

$$\left.\begin{aligned} &\lim M = 1 \quad \text{at } t \to 0 \\ &\lim M = \Pi \quad \text{at } t \to \infty \end{aligned}\right\} \tag{252}$$

It is assumed that a continuous spectrum of relaxation times τ_i exists, that the condition $t \to 0$ corresponds to a time t that is much less than the shortest relaxation time, and that the condition $t \to \infty$ corresponds to a time that t is much greater than the longest relaxation time.

Similar correlations can be obtained[31] for the mean relaxation times of different types of processes such as (a)–(d). The ideas stated here (in a rather simplified version) were initially formulated simultaneously by Moynihan and Gupta[251] and by Berg and Cooper.[254] Later these methods were developed in detail in matrix form by Moynihan[252] and by Lesikar and Moynihan,[251] whose works I used for explicating the idea. Due to the efforts of Moynihan and his colleagues, the problem of interrelations among the relaxational functions and the plurality of structural parameters has obtained an elegant lucidity. For this reason, Relation (251) may be named the Moynihan relation.

The physical sense of Correlations (251) and (252) is very simple. Earlier, when discussing the influence of the plurality of independent order parameters on the Π value, we imagined a speculative experiment at T_g involving the transition from the equilibrium liquid state to that of glass with completely frozen-in order parameters. In this case the value of Π was defined by the contributions of all the order parameters to the ΔC_p, $\Delta\kappa_T$, and $\Delta\alpha$ increments. In relaxation experiments, at the transition from glass to liquid, the complete change of these coefficients is possible only at the complete attainment of the equilibrium state with respect to all order parameters, i.e., at $t \rightarrow \infty$. Therefore, only for this case may the relation M, where the relaxational functions are in inverse proportion to the increments of thermodynamic coefficients, coincide with the value of the Prigogine-Defay Ratio.

For incomplete relaxation ($0 < t < \infty$), the inequality $M < \Pi$ means that the increment of any of the three coefficients (ΔC_p, $\Delta\kappa_T$, or $\Delta\alpha$) has not reached its maximum possible value. Here, most probably ΔC_p and/or $\Delta\kappa_T$ relax slower than $\Delta\alpha$ does. The limiting case of $M \rightarrow 1$ at $t \rightarrow 0$ corresponds to the state when relaxation has not yet begun and the initial state of glass has not changed.

3. Illustrations of the Multiplicity of Order Parameters Based on Relaxation Experiments

The above relations hold only in cases where the deviations from equilibrium are not large and the linear approximation is appropriate. In this case, positive and negative stimulating deviations of thermodynamic parameters must give the same results for the relaxation function. It has been well known for a long time that at great deviations from the equilibrium the relaxation proceeds more slowly if the fictive temperature increases in this process, and more rapidly if it decreases, even when the absolute values of the deviation from the equilibrium are equal (see, e.g., Reference 242). This manifests itself, in particular, in the dependence of the index of the stretched exponent (214) vs. the sign of the deviation. However, as confirmed by the recent adiabatic calorimetry investigations of enthalpy relaxation by Fujimori et al.,[255] in propylene glycole and glycerol near their T_g, $|\delta T| \rightarrow 0$ from both sides, the relaxation function approaches a constant value, and β assumes the value specific only for the substance and the temperature.

The analysis of relaxation functions is best done over a wide range of temporal changes and a wide range of states (thermodynamic temperature). A dynamic method of heat capacity measurements at high frequencies (in the range of $1–10^8$ Hz) developed specially for this purpose enabled Birge et al.[256] to compare the characteristics of enthalpy and density relaxation spectra (acoustic measurements). The system was glycerol at 270–390 K (the region of T_g and above). It appeared that the mean relaxation times of both processes are close to each other; the values of β are 0.65 ± 0.05 for the heat capacity relaxation and 0.60 ± 0.05 for the volume relaxation.* The relaxation times obtained for glycerol from different experiments (dielectric measurements, digital correlation spectroscopy, Brillouin scattering, and the above experiment) agree over a wide temperature range, although dielectric relaxation yields a narrower spectrum than other types of relaxation. This shows that the mean relaxation time is not a characteristic criterion for discrimination of the kind of relaxing

* The corrections for the transition from adiabatic values to isothermal ones are introduced.

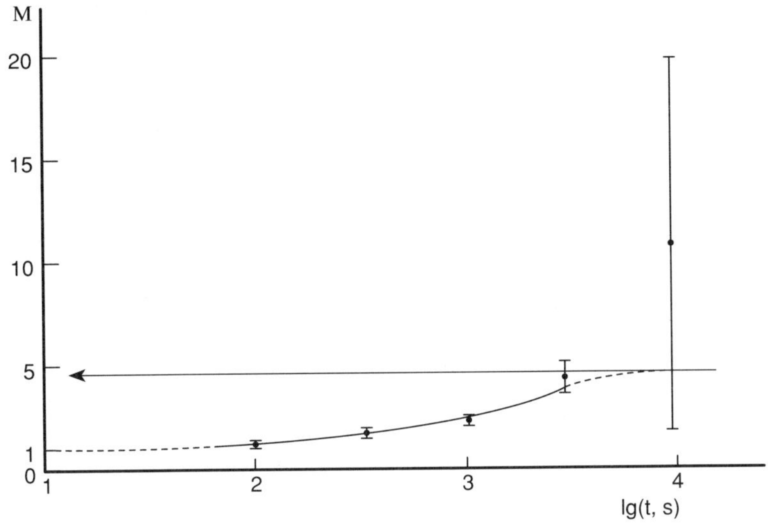

FIGURE 30. Dependence of the Moynihan ratio value (M) on the logarithm of the experiment time. The arrow points to the value of the Prigogine-Defay Ratio (Π) for B_2O_3, 543.3 K. (according to numerical values given by Moynihan et al.[31,251]).

structure parameters. Of greatest interest is the parameter β or the combination of β and ⟨τ⟩. The similarity of relaxation times and their temperature dependencies may be determined by the resemblance of the potential barriers at molecular rearrangements (i.e., by the similarity of the mechanisms as a whole). Thus, for example, the activation energies for volume and shear viscosity are practically the same, although the processes differ by their eventual *result* (i.e., the relaxation of volume and the relaxation of shear stresses, which does not change the volume; see Goldstein[181]).

In view of the above circumstances, results of the treatment of several associated relaxation experiments reported long ago by Moynihan and Lesikar[31,252] are especially important. For desiccated B_2O_3 at 543.3 K, the relaxation functions ϕ_{ST}, ϕ_{Vp}, and ϕ_{VT} are respectively characterized by the following combinations of parameters β, $\tau_{K(QX)}$, and $\langle\tau_{QX}\rangle$ in seconds: (0.65; $2.4 \cdot 10^3$; $3.3 \cdot 10^3$), (0.60; $1.6 \cdot 10^3$; $2.4 \cdot 10^3$), (0.82; $5.7 \cdot 10^3$; $6.4 \cdot 10^3$).* The dependence of the Moynihan ratio (M) on the logarithm of the experiment time is shown in Figure 30. The dispersion of M values is defined by the absolute error in the determination of each ϕ_{QX} function (± 0.02) for every time of experiment t. At $t \to 0$, $M \to 1$ and at increasing t, M falls into the region of values close to Π. This conforms with the theory. The minute course of the M change at $t \to \infty$ still presents an unsolved experimental problem.

In the above case it is possible to compare the three relaxation functions (the function ϕ_{Sp} has not been examined thus far and is assumed to be a

* Note that for significant deviations from equilibrium, the methods for the calculation of relaxation functions correctly included the relaxation nonlinearity.

theoretically obvious equality $\phi_{Sp} = \phi_{VT}$). For other substances for which it is known experimentally that $\Pi>1$, Moynihan[252] presents examples of $\phi_{VT} \neq \phi_{ST}$ or $\phi_{Vp} \neq \phi_{ST}$ inequalities, which must exist according to the theory,* as well as other examples of such inequalities when the value of Π is unknown.

The above facts support the necessity of using nonequilibrium thermodynamics methods (which consider the plurality of the relaxing structure parameters) to describe the relaxation properties of glass. The relaxation functions defined for thermodynamic parameters stimulating relaxation and those defined for relaxing properties differ from each other. The latter is manifested in unessential distinctions of the relaxation function parameters (such as the stretched exponent), which, however, should not be ignored. Volume relaxation proceeds more rapidly than that of entropy.**

4. Relaxation of Fictive Pressure Accompanying Relaxation of Fictive Temperature

At melt quenching, its structure is determined not only by the fictive temperature T_f but also well by the fictive pressure p_f. The latter corresponds to the pressure as an equal parameter of the initial state. This parameter is specified by the conditions of experiment. During the relaxation process initialized by a temperature jump δT, structure relaxation takes place, and $T_f \to T$ when $T = T_f + \delta T$. The answer to the question of what happened to the fictive pressure in that process was given recently by Gupta.[250] Naturally, if pressure p remains constant in such an experiment, p_f is equal to that pressure at both the beginning and the end of the experiment. Gupta considered the rates of T_f and p_f relaxation defined by the relations

$$-\frac{\partial T_f}{\partial t} = A_{11}(T_f - T) + A_{12}(p_f - p) \tag{253}$$

$$-\frac{\partial p_f}{\partial t} = A_{21}(T_f - T) + A_{22}(p_f - p) \tag{254}$$

$$-\frac{\partial T_f}{\partial t} = -\left[L_{TT}\left(\frac{\partial G}{\partial T_f}\right) + L_{Tp}\left(\frac{\partial G}{\partial p_f}\right)\right] \tag{255}$$

* $0.4Ca(NO_3)_2 \cdot 0.6KNO_3$, polystyrene, and polyvinyl acetate.

** The difference between the volume and the entropy "relaxation paths" can be understood, although not strictly, by an illustrative example. Let a number of fishes lie on a plane in disorder. Using a ruler one can easily, in several moves, transport them into a certain outlined square (the analogy of volume relaxation). However, in order to arrange the fishes in a specified order within the square, one would have to transfer almost every one by separate moves. This is the analogy of entropy relaxation. Obviously entropy relaxation requires more time than volume relaxation.

$$-\frac{\partial p_f}{\partial t} = -\left[L_{Tp}\left(\frac{\partial G}{\partial T_f}\right) + L_{pp}\left(\frac{\partial G}{\partial p_f}\right)\right] \tag{256}$$

The relaxational coefficients A are similar to the reciprocal relaxation times in Relation (231), which holds for the single-parameter approximation. However, each of them corresponds either to the same combination of differing order parameters responsible for the fictive temperature relaxation (A_{11}) or the fictive pressure relaxation (A_{22}), or to their combination (A_{21} and A_{12}). The coefficients L are similar to the kinetic coefficients γ in Equation (229) and depend on T and p. They differ in an analogous way. The coefficients A and L are related with each other. The [A] and [L] matrices are defined positively. The matrix of the relaxation coefficients is

$$[A] = [L][\Delta G''] \tag{257}$$

where $[\Delta G'']$, the matrix formed by the increments of the second derivatives of the free energy (i.e., by the increments of the thermodynamic coefficients that define the Prigogine-Defay Ratio), must be defined positively. From this condition Gupta derived an important expression:

$$\delta V^{(t)} = \left(\frac{TV^{(e)}\Delta\alpha}{\Delta C_p}\right)\delta S^{(t)} - \frac{(\Pi - 1)T\left(V^{(e)}\Delta\alpha\right)^2}{\Delta C_p}\left(p_f - p\right)^{(t)} \tag{258}$$

This correlates current with time deviations from equilibrium values for the volume $\delta V^{(t)}$ and the entropy $\delta S^{(t)}$ with the values of total jumps of thermodynamic coefficients, with the value of the Prigogine-Defay Ratio, and with the current deviations of the fictive pressure from the exterior pressure $(p_f - p)^{(t)}$, as well as with $(T_f - T)^t$. $V^{(e)}$ is the equilibrium volume; $\delta V^{(t)}$ and $\delta S^{(t)}$, which are described in terms of fictive temperature, are bound by linear dependence only when $\Pi = 1$, i.e., the only parameter is that of order.

In the general process of fictive temperature relaxation, $(p_f - p)$ is nonzero, though at the beginning and at the end of the process $p_f = p$. The sign of the deviation $(p_f - p)$ depends upon that of $\Delta\alpha$ because ΔC_p is always positive (Figure 31). At $\Delta\alpha = 0$ and $L_{Tp} = 0$, $(p_f - p)$ might be zero throughout the relaxation. If $\Delta\alpha < 0$, $L_{Tp} < 0$, and if $\Delta\alpha > 0$, $L_{Tp} > 0$. Kinetic restrictions must be imposed on the increments of the thermodynamic coefficients, that is,

$$\Delta\alpha \cdot TVA_{21} < A_{11} \cdot \Delta C_p \tag{259}$$

(the relaxation is caused by a T jump) and

$$\Delta\alpha \cdot A_{12} < A_{22} \cdot \Delta\kappa_T \tag{260}$$

(the relaxation is caused by a δp jump). These inequalities correspond to those required by the Moynihan ratio for the relaxation functions. The Gupta theory

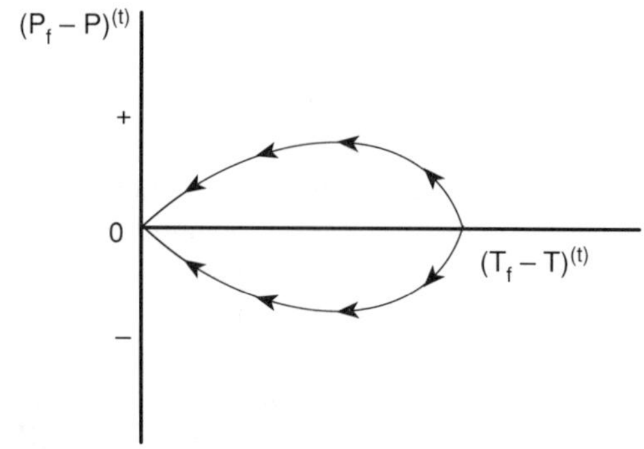

FIGURE 31. Variation of $(p_f - p)^{(t)}$ vs. the variation $T_f - T$ according to Equation (258) for different signs of $\Delta\alpha$. The upper curve is for $\Delta\alpha > 0$ whereas the lower one is for $\Delta\alpha < 0$. In Equation (258) $\delta S^{(t)}$ and $\delta V^{(t)}$ are linear functions of $(T_f - T)^{(t)}$ and of $(P_f - P)^t$.

makes the kinetic causes of the memory effect in the crossover experiments quite obvious. It reveals many possibilities for the scientific basis of new types of tempering processes.

The physical sense of Gupta's theory also is clear. If volume relaxation proceeds more rapidly than that of entropy, structural disparity appears between the entropy and the volume that exists at every time moment. The fictive pressure of such a state differs from the pressure parameter of the initial state. Depending on the sign of $\Delta\alpha$, the entropy relaxation proceeds as if under conditions of forced compression or expansion. The difference between the fictive pressure and the thermodynamic pressure is generated only in the relaxation process. The difference is defined by the degree of conformity of the current volume with the volume and entropy the system would have possessed under conditions of complete equilibrium.

The results in this chapter assure that the relaxational transition of glass to the liquid state is defined by kinetically different processes. Kinetic differences can be described in terms both of fictive temperature differences for different properties and of different groups of structure ordering parameters. The result is that structural changes have different effects on different properties.* This inevitably leads to the condition that glass genesis for gradual cooling must take place in a fairly wide temperature range. The interpretation of the glass transition in terms of purely mechanical relaxation is only a convenient means, making such a process explicitly speculative. For a mechanical glass transition only very elementary structural changes are frozen in. More complicated changes can be frozen in at different levels of completeness at higher temperatures. A similar point of view on the difference between mechanical and structural glass transitions was proposed in the 1950s by Bartenev (see, e.g., Reference 257).

* The complex character of the chemical structure of glass (covalent and van der Waals bonds, polar groups like quadrupoles, etc.) and the set of activation energies for different processes result in T_g values obtained by thermal expansion data differing from those obtained by conductivity data.[105]

VI Viscosity Theories of Glassforming Melts and Glasses

A. SPECIFIC FEATURES OF THE THEORETICAL APPROXIMATIONS

Relaxation of glass properties in a wide temperature interval has two main characteristic features. The relaxation is nonexponential vs. time and is always described by a stretched exponential law (214), where the exponent index β increases and approaches 1 with the increases. The relaxation is nonexponential vs. temperature and in the majority of cases over a wide temperature range the relaxation time τ_K does not conform to the Arrhenius law,

$$\tau_K = \tau_0 \exp(b/T) \tag{260}$$

The values of τ_0 and b change with temperature in *approximately* the same way as the viscosity parameters do. There is presently a tendency to connect these features, especially the latter one, with the same circumstances, giving rise to the peculiar temperature dependence of the viscosity.

An analysis of viscous flow theories is necessary to understand the possible causes of these facts. It should be added that viscous flow theories use some thermodynamic values which are typical of supercooled liquids. No longer can the viscosity coefficient for liquids be presented in terms of classical molecular-kinetic models as multiparameter relations where each parameter could be regarded as defined exactly (see, e.g. Born and Green[258]). One reason to reject such an approach concerning associated liquids (which is well-grounded in practice only for gases) appeared as long ago as the 1930s when the cooperative nature of atomic or molecular displacements in condensed systems (see, e.g., Barrer[259]) was recognized. The modern concept of the cooperative nature implies that the displacement of a particle from one equilibrium position to another requires significant excitation in its surroundings.

If there are directed chemical bonds in glass melts that define comparatively low coordination numbers, the translational changes of atomic sites can be compared with the ability of atoms to vary their coordination, as the bonds must be destroyed. Recalling the words of Bernal,[260] "The lower coordination produced by the existence of bonds will effect both the thermal properties and the viscosity of liquid. Change of coordination will here be associated with greater energy and the result will be a much larger configurational specific heat and correspondingly larger viscosity; there will also be, corresponding to this, a far greater tendency to form glasses … a glass being defined as a supercooled liquid in which the viscosity has become so high that the configuration remains fixed" (p. 39). Later we will see that the supposed relations are really a bit different, but the statement of the problem was (rather perspicaciously) formulated by Bernal.

The description of such complex processes necessarily requires that some information "blocks" contained in latent form such as configurational entropy, free volume, etc. be used in the theory. Any theory of this kind can only be approximate. The advantages of such theories can be revealed not only by the strictness of the simplifications used and by the degree of conformity with experiment, but also by the number of arbitrarily selected constants and universality of their values. Such theoretical approaches are now most popular and proving themselves.

In this book we cannot present an historical analysis of total variety of the ideas generated, so we confine ourselves to the most essential ones. In most cases, however, one should remember that almost every modern approach in viscosity theory has a deeply rooted prehistory and sometimes quite involuntarily invokes long-forgotten but essentially correct ideas. Regarding the significance of previous theories of viscous flow in one of the latest glass transition theories (the Mode Coupling Theory), the reader is referred to the review by Nemilov.[261]

B. PECULIARITIES OF THE VISCOSITY DEPENDENCE ON TEMPERATURE AND MELT STRUCTURE

Unless specified otherwise, Newtonian viscosity of glasses is implied because over a wide range of readily attainable shear stresses the Newton Law holds strictly. Figure 32 presents the overall pattern of changes in the viscosity of glass and its melt vs. reciprocal temperature. Below T_g (region 1), the viscosity varies exponentially, as

$$\eta = \eta_{01} \exp(E_1 / RT) \tag{261}$$

The value of T_g, the viscosity at T_g, and E_1 depend upon the cooling rate. With decreasing the latter, $\lg\eta$ at T_g and E_1 increase. The possibility of reliable

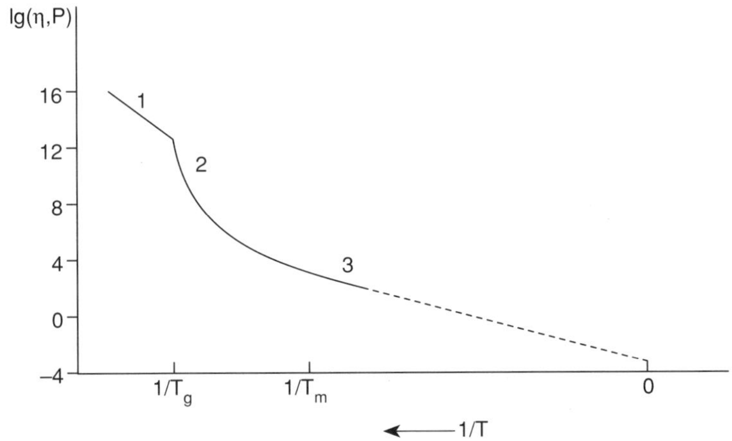

FIGURE 32. Overall pattern of changes in the viscosity of glass and glassforming liquid over a broad temperature range. For an explanation, see text. (From Nemilov, S.V., *Sov. J. Glass Phys. Chem.* (Engl. Transl. by Interperiodica Publ.), 18, 341, 1992. With permission.)

measurement of the viscosity in region 1 is defined by the significant remoteness between glass and liquid shear relaxation spectra (see Hopkins and Kurkjian[262]). Measurements of nonequilibrium states and the investigation of the kinetics of viscosity stabilization are possible (Mazurin et al.[263–265]). The viscosity remains Newtonian. Above T_g, the melt viscosity falls more abruptly (region 2); this fall is also exponential in most cases, and described by

$$\eta = \eta_{02} \exp(E_2 / RT) \tag{262}$$

where $E_2 \gg E_1$. In many cases the viscosity in the region between T_g and temperatures corresponding to ~10^8–10^{10} P is better described by the empirical FVT relation (169)

$$\eta = A \exp[B/(T - T_0)] \tag{263}$$

whose validity, as well as A, B, and T_0 values, depends on the length of the approximation interval. To predict whether the exponential or the FVT law holds is impossible beforehand. Even for glasses of the same chemical group but differing in composition by only several components percent, the dependence can change from (262) to (263).*

Near or above the melting point, an exponential dependence (region 3) is also often observed,

$$\eta = \eta_{03} \exp(E_3 / RT) \tag{264}$$

* See, e.g., the work by Nemilov et al.[266]

where $E_3 \ll E_2$ and $E_3 \lesssim E_1$. In these cases, the logarithm of the preexponential factor corresponding to region 3, $\lg\eta_{03}$, is usually in the range of -3.5 ± 1 (viscosity in poises, P). Such a preexponential value is generally typical for the viscosity of the overwhelming majority of liquids (Barrer[267]).

Very seldom can the entire curve of a glass melt viscosity from ~10^{15} to 10 P be described by the FVT equation (263) within experimental error. Such a curve is that of the viscosity standard N 710 (NBS USA).[268] However, in most cases, the FVT approximation (263) for glasses overstates the experimental values for viscosities of more than 10^{13}P, (for the latter the experimental error is $\Delta\lg\eta \sim \pm 0.03$). At lower viscosities, this discrepancy becomes sign-altering.[269] When the constants in Equation (263) have been selected for the best fit with the experiment in the low viscosities region (as Uhlmann has shown[270] by the example of α-phenyl, *o*-terphenyl and salol), at viscosities ~10^{14} P. Equation (263) yields values about five orders larger than experiment. This contrasts with the widespread conviction that the FVT law is supposedly exact and universal.

In polymers rheology the Williams-Landel-Ferry (WLF) law,

$$\lg\left(\tau/\tau_g\right) = -C_1\left(T - T_g\right)/\left(C_2 + T - T_g\right) \tag{265}$$

is widely used which correlates the relaxation times at T and T_g (τ and τ_g). It is most frequently valid for a ($T - T_g$) interval of about 50 to 80 K. Many existing theories can be reduced to the dependencies of either FVT or WLF types.

Considering the influence of the glass nature on temperature dependence of the viscosity, it is convenient to use a reduced coordinate system and to examine the dependence of viscosity on the ratio T_g/T (T_g being the temperature at $\eta = 10^{13}$ P). The method was proposed by Oldekop.[271] Obviously, at $T_g/T = 1$, the curves for all glasses converge in one point, whereas at $T_g/T \to 0$ all the curves proceed to the limit $\lg\eta_{03} \sim -3.5 \pm 1$. The divergence between the curves (different degree of curvature) in the intermediate region is determined by the value of the derivative $\partial\lg\eta/\partial(T_g/T)$, i.e., how drastically the viscosity changes with temperature near T_g. This is equivalent to the divergence of the preexponent η_{02} and η_{03} logarithms for glass transition range* and for the liquid range [Equations (262) and (264)].

Nemilov,[272–274] as long ago as 1964, showed that smallest difference between the values of η_{02} and η_{03} was observed for glasses formed by tetrahedrally linked structural polyhedra (SiO_2, GeO_2, BeF_2, $GeSe_2$, etc.). For such glasses, the slope change at the transition from region 2 to region 3 is very slight or absent. Glasses formed by chains (Se, polyisobutilene, sodium metaphosphate, etc.), molecular glasses, and those that are molecular with hydrogen bonds (glycerol, glycose, and others), as well as those with highly ionic bonding, having low (salt-like) viscosity in their melts (as $La_2O_3 \cdot 3B_2O_3$,

* If the viscosity is nonexponential near T_g, the logarithm of η_{02} is determined from the equation of the curve tangent.

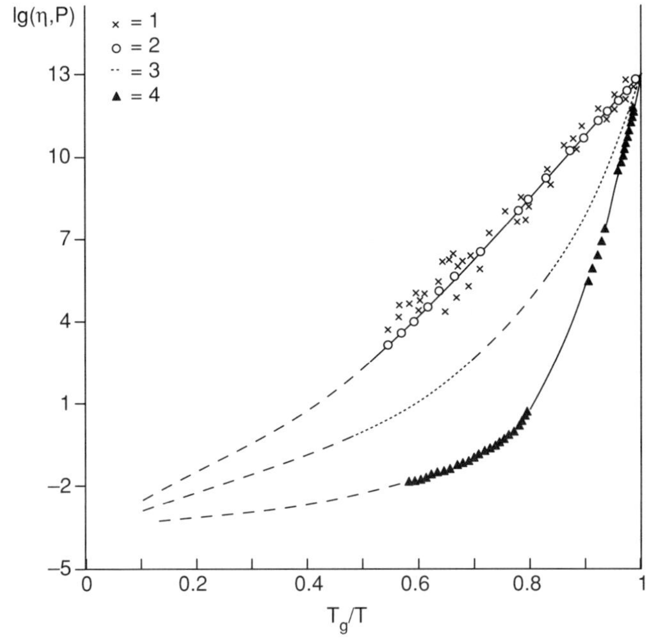

FIGURE 33. The viscosity logarithm, lg(η,P), as a function of the reduced reciprocal temperature, T_g/T, for representative glasses of two limiting classes. SiO_2 (1) and BeF_2 (2) are "long" glasses, Se (3) and *o*-terphenyl (4) are "short" glasses. For references to the experimental data, see Nemilov.[275]

$2PbO \cdot SiO_2$, $2PbO \cdot B_2O_3$, etc.), are characterized by the maximum concavity of the viscosity curves and by the maximum divergence between $\lg\eta_{03}$ and $\lg\eta_{02}$. Glasses whose structure is formed by polyhedra having three bridging bonds (as $Na_2O \cdot 2SiO_2$, $K_2O \cdot 2SiO_2$, As_2Se_3, As_2S_3, B_2O_3, etc.) lie between the first two groups. This classification, according to the principles of spatial connectivity of glassy network and chemical bonding, was made based on an analysis of the activation entropies in the T_g-region which, according to the conventional designation, are calculated as

$$\Delta S_\eta^{\ddagger} = R\ln\left(\eta_{03} / \eta_{02}\right) \tag{266}$$

For the first group of glasses, $\Delta S_\eta^{\ddagger}$ values vary from 40 to 170; for the second they are over 800; and for the intermediate ones the values fall into the interval of 290 to 590 $J \cdot K^{-1}mol^{-1}$. In Russian and German literature on technology the two limiting groups are named "long" and "short" glasses, respectively. Nemilov[272] interpreted this correlation based on an analysis of the heat capacity (ΔC_p). An example of such a comparison of glasses is plotted in Figure 33 in the coordinates $\lg\eta - T_g/T$. Laughlin and Uhlmann[276] and, especially, Angell et al.,[277–282] have used this type of graphical presentation of data. Angell et al., apparently unaware of Nemilov's works, actually confirmed the above classification. The "longest" of the known glasses is vitreous SiO_2, while the "shortest" are those based on ZrF_4 and the glass of composition $0.4Ca(NO_3)_2 \cdot 0.6KNO_3$.[277–282] Angell called the first group of glasses "strong glasses" and the second group "fragile glasses". The essential meaning of these terms is that at transition from the first glasses to the second, the melt structure gains more possibilities for configurational changes with increasing temperature, i.e., loses its "strength." Every viscosity theory must in principle explain the above correlation.

C. FREE VOLUME THEORY

The simplest models of the liquid state consider the existence of molecular-size holes in liquids from which the equation of state for a liquid is derived. Treatments of such models by Frenkel, Eyring, and others are beyond the scope of this book and the reader is referred to appropriate surveys (e.g., Reference 283). We also do not consider empirical correlations between the free volume and the viscosity for nonassociated and low-associated liquids originated by Batschinski.[284] Interest in this concept related to the glass transition appeared in the work by Doolittle,[285] who showed that the viscosity of simple liquid hydrocarbons could be described by the equation

$$\eta = A_0 \exp[\gamma V_0 / V_f] \tag{267}$$

where V_0 is the volume at zero K, $V_f = V - V_0$ (V_f is the free volume, V is the volume at the temperature T), and the coefficient γ is about 1. The Doolittle equation is more frequently given as

$$\eta = \eta_0 \exp(b/f) \tag{268}$$

where f is the free volume fraction defined by

$$f = (V - V_0)/V \tag{269}$$

Williams, Landel, and Ferry[286] showed that in the transition temperature range the relaxation time for a large number of glassforming substances was described by Equation (267), provided that the free volume was defined as

$$V_g = V[0.025 + \Delta\alpha(T - T_g)] \tag{270}$$

where V_g is the volume at T_g, 0.025 is an empirical constant determining the free volume fraction at T_g, and $\Delta\alpha = \alpha_l - \alpha_g$ is the thermal expansion increment at T_g due to the free volume increasing. Then

$$f = f_g + \Delta\alpha(T - T_g) \tag{271}$$

This relation leads to the WIF relation (265), where the C_1 and C_2 constants are related to the coefficient b of Equation (268) and to $\Delta\alpha$ and f_g via

$$C_1 = b/2.303 f_g \quad C_2 = f_g / \Delta\alpha \tag{272}$$

The value of f_g for a wide range of substances of both organic and inorganic nature varies within a fairly narrow range of 0.02 to 0.03; the value of C_1 is also constant. Thus the criterion of a glass transition is that the free volume fraction reach its critical value of ~0.025. Other details of this theory can be found in the book by Sanditov and Bartenev.[283]

Cohen and Turnbull[287] proposed a statistical basis for the Doolittle dependence (267) based on the following assumptions:

1. It is possible to associate a certain local volume v of a molecular scale with each molecule. This volume is not identical to a molecule's own volume but is defined as a part of system's volume per molecule. The volume v is similar to Voronoi's polyhedron.
2. The free volume is the volume that exceeds a certain critical value v_c.
3. To redistribute the free volume among molecules, no local energy is expended. The free volume $v_f = v - v_c$ of each molecule has fluctuations and its distribution is continuous within the free volume of all N particles ($V_f = Nv_f$).
4. Molecular transport (molecule jumps) is possible only when the molecule free volume is greater than a critical value of the order of molecular size ($v_f > v_f^*$).

Applying usual combinatorics for the expression of the entropy due to the free volume distribution among N molecules, Cohen and Turnbull derived an equilibrium law of the free volume distribution from the condition of the entropy maximum. Hence they found the fraction of molecules having $v_f > v_f^*$, the probability of molecular jumps, and an expression for viscosity

$$\eta = A' \exp\left(N v_f^* / V_f\right) \tag{273}$$

which agrees with the Doolittle equation ($\gamma V_0 \simeq N v_f^*$). The A' value, in addition to other factors, includes the square of molecule diameter, which correlates with V_0. There is no temperature in an explicit form in the Doolittle relation. Whereas in the Maxwell-Boltzmann statistics the exponent is defined by the relation of the energy of a particle to the system temperature, which characterized the average thermal energy, here the free volume fluctuations are similar to those of thermal energy.

The FVT equation (263) from Relation (267) may also be derived if

$$V_f / V = \Delta\alpha\left(T - T_0\right) \tag{274}$$

is assumed in (270), where T_0 is a temperature analogous to the Kauzmann temperature T_K. At T_0, $V_f = 0$; this point should be defined by extrapolating the volume into the low-temperature region. The theory of the free volume in this simple version also presumes the dependence of the free volume fraction on pressure,[287]

$$f \equiv V_f / V = f_g + \Delta\alpha\left(T - T_g\right) - \Delta\kappa_T p \tag{275}$$

Here $\Delta\kappa_T$ is the increment of the compressibility at T_g, which, as well as $\Delta\alpha$, is attributed only to the free volume. The latter is a generalized structural parameter. Kovacs[288] showed that in this case the baric dependence of T_g (f freezing) must be expressed as

$$dT_g / dp = \Delta\kappa_T / \Delta\alpha \tag{276}$$

This dependence disagrees with experiment (Goldstein[289]), as it appears that

$$dT_g / dp < \left(\Delta\kappa_T / \Delta\alpha\right) \tag{277}$$

Apparently the reason is the same as that explaining the difference of the Prigogine-Defay Ratio from 1, namely

$$\left(T_g V \Delta\alpha / \Delta C_p\right) < \left(\Delta\kappa_T / \Delta\alpha\right) \tag{278}$$

According to Goldstein,[289] the relation

$$dT_g / dp = T_g V \Delta\alpha / \Delta C_p \tag{279}$$

being more justified experimentally, points out that the entropy or the enthalpy, but not the free volume, define the glass transition.

Kumar[290] used a simple version of this theory to determine the size of kinetically mobile units in borate and silicate glass melts using viscosity and density data. According to his results, these are large groups containing about two or more elementary structural units ($[Si_2O_7]$, $[Si_6O_{18}]$, or $[B_4O_7]$).

Macedo and Litovitz[269,291] proposed a hybrid equation for viscosity,

$$\eta = \eta_0 \exp\left[\left(E_V^* / RT\right) + \gamma\left(V_0 / V_f\right)\right] \tag{280}$$

which is based on the postulate that for a molecular jump not only the free volume but also the simultaneous overcoming of the potential barrier is necessary. This equation is in better agreement with experiment for a wide range of viscosities. Besides, unlike the simplest version of the free volume theory, it renders correct viscosity dependencies on pressure and on temperature simultaneously. In these authors' theory the free volume is approximated by the two-states model,

$$V_f = \Delta V \left\{1 + \exp\left[(\Delta H - T\Delta S) / RT\right]\right\}^{-1} \tag{281}$$

where ΔV, ΔH, and ΔS are the differences in the volume, enthalpy, and entropy of these states. Thus the number of constants to be determined increases. The disparity between the simplest theory and experiments on glass viscosity at constant volume sustained by compressing pressure was also pointed out by Goldstein.[181] He emphasized that in viscosity theory the question of the mechanism of local transformation of macroscopically specified stresses into microscopic stresses resulting in concrete relaxation and mass transport remains vague. Mooney[292] attempted to consider this problem in connection with the free volume theory. He showed that fluctuational changes of free volume were caused by fluctuations of thermal vibrations and there led inevitably to instantaneous elastic deformations of the medium. The elastic deformations region includes many molecules. The transition to Newtonian flow necessarily proceeds as an activated process in which critical stresses are to be calculated from instantaneous elastic characteristics of liquid. Therefore the activation energy must be proportional to the energy of elastic distortions, the latter being proportional to the critical increment of the local volume at which the local elasticity disappears and viscous shear occurs.

The free volume theory has been developed further by excluding the concept of a potential energy barrier although an interpretation of its parameters exactly in such a context exists.[283]

Cohen and Grest[293–295] essentially changed the free volume theory and brought it nearer to the conceptual thermodynamic principles of the glass transition theory. Their model is based on the theory of liquid-like clusters. Following Kirkwood's approach, the authors defined the free energy of the system through a cellular free energy f_i depending only on the cellular volume,

$$G = N\int P(v)\left[f(v) + k_B T \ln P(v)\right]dv - TS_{com} \tag{282}$$

where P(v) is the probability that a cell has the volume v and S_{com} is the communal entropy defined by free exchange of cell sites involved in a liquid-like cluster. The theory includes the postulate that a local free energy depending on volume varies according to different laws at small and at large values of that volume. Near the minimum,

$$f(v) = f_0 + \frac{1}{2}k\left(v - v_0\right)^2 \tag{283}$$

for $v < v_c$. For $v > v_c$, the other law holds;

$$f(v) = f_0 + \frac{1}{2}k\left(v_c - v_0\right)^2 + \zeta\left(v - v_c\right) \tag{284}$$

This corresponds to the linear dependence of the function f(v) on v. The cells corresponding to Equation (283) are regarded as solid-like, whereas those corresponding to Equation (284) are regarded as liquid-like. The fraction of liquid-like cells is defined as

$$p = \int_{v_c}^{\infty} P(v)dv \tag{285}$$

Liquid-like and solid-like cell exchange is hindered by requiring large expenditures of free energy whereas exchange among liquid-like cells proceeds freely (a liquid-like cluster). The ability to exchange freely is defined by the coordination number of a liquid-like cell z, i.e., by the condition of how many liquid-like cells surround the given cell of the same type. The coefficient ζ breaks into two components, ζ_0 and ζ_1. ζ_0 is constant, and ζ_1 depends on the state of the cell surroundings,

$$\zeta_1 = \frac{k_B T_1}{v_a + \bar{v}_f} \tag{286}$$

and

$$\zeta_1 = \frac{k_B T_1}{v_a + \bar{v}_f} + \zeta_0 \tag{287}$$

where T_1, v_a, and ζ_0 are constants and $\bar{v}_f$ is the mean free volume per cell. The latter is defined as

$$\bar{v}_f = \frac{\int_{v_c}^{\infty} (v - v_c)P(v)dv}{\int_{v_0}^{\infty} P(v)dv} \tag{288}$$

Here P(v) is the same as in Formula (285). After other simplifications, the expression for $\bar{v}_f$ in terms of the above characteristics becomes

$$\bar{v}_f = \frac{k_B}{2\zeta_0}\left\{T - T_0 + \left[(T - T_0)^2 + \frac{4v_a\zeta_0 T}{k_B}\right]^{1/2}\right\} \tag{289}$$

where

$$k_B T_0 = kT_1 + v_a\zeta_0 \tag{290}$$

The total free volume of an entire liquid-like cluster for nonactivated diffusion must be greater than the critical value v_m.

The communal entropy connected with the accessibility of the entire configurational volume within the limits of finite or infinite clusters is defined by the expression

$$S_{com} = Nk_B pa_Z(p)\left\{\sum_{\nu=\nu_m}^{\infty} \nu C_{vz}(p)\ln\left[(1-\nu)\overline{v}_s\right] + P_z(p)\ln\left[Npa_z(p)P_z(p)\overline{v}_s\right]\right\} \quad (291)$$

Here Np is the total number of liquid-like cells in the system ($N \cdot p \equiv \int_{v_c}^{\infty} P(v)dv$) and p is their fraction, $Npa_z(p)$ is the number of liquid-like cells in the cluster, C_{vz} is the characteristic of cluster distribution by their sizes, ν is the reduced value of the cluster free volume, ν_m is the reduced value of the cluster critical free volume ($\nu_m \equiv v_m/\overline{v}_f$), and $\overline{v}_s$ is the mean configurational volume of a liquid-like cell. The introduction of a thermodynamic criterion for equilibrium (the entropy maximum) enabled Cohen and Grest to establish that the value of p depends upon coordination number z and the cluster can be liquid-like only if $z > 2$. There exists a threshold value of cell fraction p_{cz} below which the free exchange among cells is impossible because of their small number and size and above which appear the conditions for free percolation. The percolation criterion here is a bit different from the usual one because it is defined by the thermodynamic state of the system. The value of P_z depends on p_z, on the critical value p_{cz}, and on the parameter β,

$$P_z(p) = B_z(p - p_{cz})^{\beta} \quad (292)$$

The parameter β is associated with the material structure (for a polymer constructed of segments $\beta \geq 1$; for a single molecule of small size $\beta < 1$).

The result is that the dependence of the entropy on liquid-like cluster fraction p has a step character, that is, S_{com} falls drastically when p becomes lower than p_{cz}, like a structural parameter change at a first-order phase transition. Since all the variable parameters of the theory depend on temperature, this is equivalent to a drastic fall of p occurring below T_p; below this temperature the system behaves as a solid (a glass). Leaving aside other details of the theory, we note here only that the viscosity expression derived in it has the following form

$$\lg\eta = A + 2B/\left\{T - T_0 + \left[(T - T_0)^2 + 4v_a \cdot \zeta_0 T\right]^{1/2}\right\} \quad (293)$$

where $B = v_m \cdot \zeta_0 \lg_{10} e$, v_m is the volume close to that critical for a cluster, T_0 is the temperature defined by Equation (290) by the sense close to T_K or to T_0 in the FVT equation. Fitting the four constants (A, B, ($v_a \cdot \zeta_0$), and T_0 in Equation (293) yields a perfect description of the viscosity behavior of tri-α-naphthylbenzene, *o*-terphenyl, α-phenyl-*o*-cesol, salol, B_2O_3, and $0.6KNO_3 \cdot 0.4Ca(NO_3)_2$. At high viscosities, Relation (293) has an exponential dependence on temperature and approaches the Doolittle approximation (267) or the FVT expression (263). Assuming

$$T_0 = T_0' + \frac{v_a p}{k_B} \tag{294}$$

where p is the pressure, Cohen and Grest obtained the baric dependence of the viscosity. Examination of v_m and v_a for the nitrate glass led them to the conclusion that the diffusing units are not individual ionic species but more complex formations.

This theory also revealed the possibility of describing equilibrium thermodynamic properties of a liquid (e.g., the heat capacity), of calculating the heat capacity jump at T_g, and of calculating the heat capacity and volume relaxations during glass annealing. The cause of the relaxation changes is that near p_{cz} redistribution between liquid-like and solid-like clusters results in a time dependence of the threshold value p_{cz} and the relaxation time is defined by most inhibited exchange processes between the two types of cells at the boundary between solid-like and liquid-like clusters. The mechanism of this exchange is presumably the same as the one at viscous flow,

$$\tau(T) = \tau_0 \exp\left(v_m / \bar{v}_f\right) \tag{295}$$

τ_0 is defined by the ratio of the surface S to the volume V of a liquid cluster,

$$\frac{1}{\tau_0} = \frac{1}{16} \cdot \frac{S}{V} \sqrt{\frac{3k_B T}{M}} \tag{296}$$

where M is molecule mass. Here the relaxing parameter is the fraction of liquid-like clusters at the temperature T which approaches the equilibrium value p_{eq} (T),

$$p(T) = p_{eq}(T) + \left[p^0(T) - p_{eq}(T)\right] \exp\left[-t / \tau(T)\right] \tag{297}$$

The theory also gives[296] an estimate of the stretched exponent law index β,

$$\beta = y / (x + y) \tag{298}$$

($^2/_3 < \beta < 1$) where x is the parameter determining the width of the distribution of the liquid cluster surface-to-volume ratio ($0 < x \le {}^1/_3$) and y is the parameter determining the width of the distribution of such cluster sizes ($^2/_3 \le y < 1$). The stretched exponential law appears as a result of the relaxation of a set of species differing by their relaxational characteristics.

Liquid-like clusters of small size frozen in glass bulk below T_g are localized. The theory of such states provides a qualitative description of small-scale relaxation processes in glass at low temperatures including those by tunneling.

The entropy of glass at absolute zero corresponds to that for solid-like cells as the communal entropy disappears. Here the solution of the Kauzmann Paradox is connected with the possibility or impossibility of a phase transition which must, however, be of the first but not second order.

The free volume theory in the new version described here has not yet received wide application. This theory leaves aside the question of the character of the temperature dependence of glass viscosity below T_g (glass viscosity with the frozen-in fraction of the free volume). However, it illustrates rather speculatively one way of taking into account the cooperativity of molecule movement. The cooperativity is included by the introduction of liquid-like cells fraction and by the notion of liquid-like clusters, which introduces fluidity provided there is spatial contact between the cells inside it, i.e., above the percolation threshold. At the same time, that very parameter can relax and can be regarded as a generalized structural parameter.

D. CONFIGURATIONAL ENTROPY THEORY

1. Derivation of the Basic Relation

The relaxation theory based on the configurational entropy concept was suggested by Adam and Gibbs[297] in 1965 and is widely used today. In its basic idea, the theory is closely connected with the statistical entropy theory of glass transition of polymers suggested by Gibbs and DiMarzio,[40–42] which we have already mentioned. The key idea of the theory is that the structural relaxation connected with overcoming the free energy potential barrier is considered possible only in the limits of a certain region of cooperative rearrangement. The size of this region is defined by the value of the configurational entropy S_c of a system consisting of N particles and by the smallest change in the configurational entropy of a subsystem s_c^* that is necessary for a monomer segment overcoming the potential barrier $\Delta\mu$ and that results in a configuration change.

Defining the smallest size of cooperative rearrangement region as z*, although in principle a wide set of such sizes is possible, the transition probability is given by

$$\overline{W}(T) = \sum_{z=z^*}^{\infty} A \exp\left(-\frac{\Delta\mu}{k_B T}\right)^z \tag{299}$$

This leads to the expression

$$\overline{W}(T) = \overline{A} \exp\left(-z^* \Delta\mu / k_B T\right) \tag{300}$$

The subsystem configurational entropy is expressed as

$$s_c = -\left(\partial G_c / \partial T\right)_{p,z} \tag{301}$$

where

$$G_c = -k_B T \ln \Delta_c \tag{302}$$

and Δ_c is the configurational partition function.

The configurational entropy of a total macroscopic system consisting of N subsystems is defined additively as

$$S_c = Ns_c \tag{303}$$

At the same time

$$S_c = k_B \ln W_c(U, V) \tag{304}$$

where W_c is the equilibrium number of configurations corresponding to the energy U and the volume V of the whole system. If N is Avogadro's number, corresponding to a mole of segments, then

$$s_c = k_B \ln\left(W_c^{z/N}\right) \tag{305}$$

The configurational entropy of a subsystem grows with increasing z. In this case, $W_c^{z*/N}$ is the mean number of configurations existing in a large system and formed of small subsystems of a size no less than z* critical for relaxation. The critical minimum configurational entropy of a subsystem is associated with its minimum size z* by the relation

$$s_c^* = k_B \ln\left(W_c^{z*/N}\right) \tag{306}$$

Hence we have the correlation between the macroscopic configurational entropy S_c, the minimum size z* of the cooperative rearrangement region permitting the critical value s_c^*, and the entropy s_c^*,

$$z^* = Ns_c^* / S_c \tag{307}$$

This produces an expression for the probability of relaxation process connected with configuration change,

$$\overline{W}(T) = \overline{A} \exp\left(-\Delta\mu s_c^* / k_B T S_c\right) \tag{308}$$

or

$$\overline{W}(T) = \overline{A} \exp\left(-C' / T S_c\right) \tag{309}$$

The temperature dependencies of $\overline{A}$ and $\Delta\mu$ are neglected and s_c^* must correspond to at least two possible states, that is,

$$s_c^* = k_B \ln 2 \tag{310}$$

The viscosity $\eta \sim 1/\overline{W}(T)$ and must accordingly become

$$\eta = \eta_0 \exp\left(\Delta\mu \cdot s_c^* / k_B T S_c\right) \tag{311}$$

If, according to the glass transition theory of Gibbs and DiMarzio,[40–42] we assume that there is a temperature T_2 analogous to the Kauzmann temperature T_K when $S_c = 0$, then at $T \rightarrow T_2$ the viscosity and structural relaxation time must increase endlessly and the temperature T_2 becomes unattainable. Only the most probable configuration of the whole system would correspond to this temperature, and the size of the cooperative rearrangement region z* would approach the size of the entire system. For an infinite temperature increase, $(S_c/N) \rightarrow s_c^*$ and $z^* \rightarrow 1$. Thus, a parameter is introduced into the theory that essentially depends on the temperature, namely the size of a region where configurational rearrangement can take place connected with a structure unit overcoming the potential barrier locally and with local rearrangement of the configuration.

In terms of the specified approximation, the Adam and Gibbs theory is perfectly strict and attractive by its simplicity. A more general derivation of Relation (308) was given by Mohanty.[298]

2. Experimental Verification

Relation (308) can be reduced to the form of the WLF relation (265) provided ΔC_p is constant in the range from T_2 to T. However, as Adam and Gibbs have shown,[297] no universal values of the parameters can be obtained and the parameter C_2 depends only slightly on the temperature. The parameter C_1 is defined by the relation

$$C_1 = 2.303\,\Delta\mu \cdot s_c^* / k_B \Delta C_p T_g \ln\left(T_g / T_2\right) \tag{312}$$

and can be universal only on the condition that $\Delta\mu$ is proportional to T_g and the T_g/T_2 ratio is constant (~1.3). Chang et al.[299], even before the publication of the work by Adam and Gibbs, had used the Doolittle's relation (268) presented in the way similar to Equation (311),

$$\eta = \eta_0 \exp\left(B'' / S_c\right) \tag{313}$$

Using the approximation

$$S_c = S_{cg} + \left(\Delta C_p / T_g\right)\left(T - T_g\right) \tag{314}$$

where S_{cg} is the value of S_c at T_g and ΔC_p is the heat capacity increment at T_g (both values being calculated per mole of "beads"), they had obtained

$$\left.\begin{aligned} C_1 &= B''/2.303\,S_{cg} \\ C_2 &= S_{cg}T_g/\Delta C_p \end{aligned}\right\} \tag{315}$$

Obviously, $B'' \equiv 2.303\Delta\mu \cdot S_c/k_BT_g$.

Including kinetics and calorimetry data, Chang et al.[299] showed that S_{cg} (as well as the free volume fraction at T_g) has significant dispersion, but on the average is ~2.9 $J \cdot K^{-1}$ per mole of "beads." The dispersion of C_1 values appears to be much less (±20%), whereas the dispersion of C_2 values is much more than that for C_1, from 246 for B_2O_3 and 193 for polystyrene to 28–50 K for *n*-propanol, glycerol, glucose, etc. This seems to be due to nonuniformity of the definition of which particle is the genuine "bead" in complex systems and to the ΔC_p dependence on temperature and the S_{cg} dependence on the value of T_g.

Greet and Turnbull[300] considered that among structurally similar glassforming molecular substances the configurational entropy at T_g might constitute a typical fraction value (~45%) of its value at T_m.

Mohanty[301] showed theoretically that the shift parameter in the WLF equation (265) could be correlated with the temperature dependence of the cooperative rearrangement region size, z*,

$$\lg\left(\tau/\tau_g\right) = -\left(\Delta\mu/2.3k_B\right)\left[z^*\left(T_g\right)/T_g - z^*(T)/T\right] \tag{316}$$

Relation (311) can also be reduced to the FVT equation form (269). If the configurational heat capacity above T_2 varies according to the law

$$\Delta C_p = K/T \tag{317}$$

by use of the approximated calculation of the integral for S_c, we have the law

$$\eta = \eta_0 \exp\left(\frac{DT_2}{T - T_2}\right) \tag{318}$$

This differs from the FVT law by substitution of T_0 for T_2. However complete conformity cannot be expected because the ΔC_p dependence on T can both decrease and increase. Richet[302] found for glassforming alumina-silicate minerals and simple sodium silicates that T_2 (T_K) and T_0 in the FVT equation (263) diverged by up to ±150 K for absolute values of T_0 from 300 to 800 K. For SiO_2, $T_0 = 0$, the viscosity depends exponentially on temperature, and T_K = 1150 K. Similar divergences also exist for molecular organic glasses and polymers, but we do not dwell upon them here.

Adam and Gibbs[297] showed the validity of Relation (308) in the approximation of the WLF equation (265) for relaxation of polymers, mono- and polyatomic alcohols, and glucose for temperatures exceeding T_g by not more than 60 K. The $\Delta\mu$ values appeared nearer to those for hydrogen bonds' energy than to those for van der Waals interactions. For polymers, however, $\Delta\mu$ is much lower (~20 kJ · mol^{-1}) than for glucose and alcohols (~70 kJ · mol^{-1}). Laughlin and Uhlmann,[276] for the description of α-phenyl, salol, *o*-terphenyl, and tri-α-naphtylbenzene viscosity curves, obtained the values of $\Delta\mu$ of about 340 to 1900 kJ · mol^{-1}. These numbers are ten times greater than the energies of cohesion calculated for a single bead. Hodge,[303,304] based on an analysis of enthalpy relaxation kinetics for polymethyl methacrylate, polyvinyl acetate, polystyrene, and other polymers which require that the temperature dependence of the relaxation time be specified, concluded that the $\Delta\mu$ value in the Adam-Gibbs equation qualitatively correlates well with the value of rotational energy barriers. For such a correlation it is necessary to assume $s_c^* = k_B \ln(3!)$ but not $s_c^* = k_B \ln 2$. This is explained by the fact that only a part of the ΔC_p value is caused by the contribution from cooperative transformations.

Araujo,[305] to describe the reduced viscosity values $\lg(\eta_{670}/\eta_T)$ for B_2O_3 in the range of 670 to 1560 K, obtained the value $\Delta\mu = 485$ kJ · mol^{-1} ($s_c^* = k_B \ln 2$), which is the same as the energy of B–O bond breaking. Scherer[306] found that the best approximation to the viscosity values of NBS standard glass N 710 (a soda-lime-silicate glass) in the region of 10^{15} to 10^9 P was obtained at $\Delta\mu = 420$ kJ · mol^{-1}; this value coincided with the energy of Si–O bond breaking in silica. The viscosity of SiO_2 and of glasses formed by aluminosilicate minerals melts in a wide range, according to the data of Richet,[302] is described by $\Delta\mu$ varying from 123 kJ · mol^{-1} for SiO_2 to 814 kJ · mol^{-1} for $NaAlSi_3O_8$ at $s_c^* = k_B \ln 2$. When these values have been reduced per mole of structural beads (i.e., divided by the number of bridging oxygens in one molecule), the divergence between the figures diminishes significantly (61.5 and 102 kJ · (mole beads)$^{-1}$, respectively). We can see that $\Delta\mu$ values are not always correlated with the energy characteristics of interatomic (inorganic glasses) or of intermolecular (organic glasses) interactions. This may be due to the nonuniform choice of structural bead for the S_c calculations. This aspect of the calculations is not clarified in most cases.

The question of the possibility of an exact description of the viscosity curve for a substance in a wide temperature range using configurational entropy theory should not be considered as finally resolved at the present time. Greet and Turnbull,[300] using the example of *o*-terphenyl, cast doubt on the agreement between theory and experiment because referring to experimental data in either glass transition range or melting point region gave different results. Later, Chang and Bestul[135] showed much better conformity of experimental data with calculations by the Adam-Gibbs theory when using precise heat capacity data and zero-point entropy values. But they did not clarify the calculations by the values of the parameters selected. Laughlin and Uhlmann[276] and Uhlmann,[270] for salol, α-phenyl-*o*-cresol, *o*-terphenyl, and other substances, showed by new experimental data that none of the theories, including that of Adam and Gibbs, gave exact coincidence with experiment over the entire viscosity range from ~10^{-2} to 10^{13} P. For these substances the viscosity depends exponentially on temperature in the range from 10^7 to ~10^{13} P. The dependence of lgη vs. $(TS_c)^{-1}$ is nonlinear and can be regarded as linear in only a rough approximation. For good agreement, a rather varying value of Δμ should be assumed. Good agreement of theoretically calculated data with those of experiment for aluminosilicate mineral melts, was shown by Richet,[302] for the above-mentioned parameters and the region of 10^0 to 10^{13} P.

It should be noted that verification of the configurational entropy theory, with respect to viscosity or relaxation time, is often not carried out directly by calorimetric experiment but in an approximation presuming *a priori* a law for change of heat capacity. The Adam-Gibbs equation then takes another form. In the case of constant ΔC_p it has the form

$$\eta = \eta_0 [B / T \ln(T / T_2)] \tag{319}$$

and becomes a three-parameter equation (η_0, B, and T_2 are constants). Better agreement with experiment is achieved more easily than in the case of Equation (311), which includes only the two parameters (η_0 and the product $\Delta\mu \cdot s_c^*$).

The viscosity of glass with frozen-in structure is described by the Adam-Gibbs theory very simply. If at the temperature T_f the configurational entropy $S_c(T_f)$ is being frozen, this value defines the slope of the dependence of lgη vs. 1/T (see Figure 32) with the angular coefficient of $\Delta\mu \cdot s_c^*/S_c(T_f)$. Scherer[306] found practically complete agreement of thus-defined viscosity activation energy below T_g for glass N 710 (283 kJ · mol^{-1}) with the experimental value (297 kJ · mol^{-1}), although for the S_c calculation he had to change slightly the parameters of the experimental temperature dependence of ΔC_p.

The configurational entropy theory is successfully used to describe structural relaxation. Formally, this reduces to the expression for $\Delta H^\ddagger$ in Equation (221) through the parameters of the Adam-Gibbs theory (Howell et al.[307]),

$$\frac{\Delta H^\ddagger}{N} \equiv k_B \cdot \partial \ln \eta / \partial(1/T) = \frac{\Delta\mu \cdot s_c^*}{S_c} + \frac{\Delta\mu \cdot s_c^* T}{S_c^2}\left(\frac{\partial S_c}{\partial T}\right) \tag{320}$$

For the description of nonisothermal relaxation in terms of fictive temperature, T_f is substituted for T and the factor dT_f/dT appears in the second item. Volume relaxation of silicate glass in different temperature and time regimes is described much better in this version than in the approximation of constant $\Delta H^\ddagger$ (Scherer[306]). The enthalpy relaxation in polymers was successfully described by Hodge.[303,304] However, for simplicity an approximation (319) was used. The latter author found empirically that the stretched exponent law index β increases more rapidly than linearly with the ratio T_f/T_2 (for not only polymers but also As_2S_3, nitrate glass, sodium-potassium silicate, and other glasses). When T_f/T_2 is increased from 1.4 to 2.2, the value of β increases from ~0.4 to ~0.9 due to the lack of cooperativity (the decreasing size of the rearrangement regions). The value x in Relation (221) appears connected with $\ln(T_f/T_2) \equiv L$,

$$x = L/(1+L) \tag{321}$$

This relation is useful for practical calculations.

The Adam-Gibbs theory is of great use for the comprehension of the general classification of glasses as "strong" and "fragile" (or "long" and "short"). In a series of his works, Angell[277–282] pointed out that "strong" glasses have little increase in their heat capacity at T_g (SiO_2, GeO_2, BeF_2), and as the "fragility" increases, ΔC_p grows too. Thus, for the same T_g/T values at $T > T_g$, configurational changes for the first group of glasses are much less than for the second group. The "strength" of glasses denotes their greater stability of structure with respect to configurational changes. Remembering the accepted classification,[271–275] we can say that the augmentation of glass "length" corresponds to the increase of structure stiffness with increases in the spatial connectivity of structural polyhedra (the transition from molecules and chains to layers consisting of triangles $AX_{3/2}$ and then to networks of $AX_{4/2}$). Structures formed by molecules or ions with nondirected coulomb interactions are most subject to configurational changes. This correlation can be understood only on the condition that a good theory of configurational entropy exists, which unfortunately is not the case today.

Angell[280,282], in terms of the Adam-Gibbs theory formalism and the FVT relation, as it has been already mentioned [Equation (318)], introduced the fragility parameter D, which defines viscosity in terms of a critical phenomenon,

$$\eta = \eta_0 \exp(D/\varepsilon) \tag{322}$$

where

$$\varepsilon = \left(T/T_0\right) - 1 \tag{323}$$

Obviously, with decreasing D, for close values of T/T_0, the fragility increases. For strong glasses, $D \approx 100$, whereas for fragile glasses (*o*-terphenyl, nitrate glass, etc.), $D = 5$. In the glass transition region ($10^{15} - 10^9$ P), the Arrhenius dependence of viscosity, which holds for "strong" glasses (SiO_2, GeO_2, etc.), gradually ceases with decreasing D, being manifested only for some glasses (As_2O_3, B_2O_3), and then often appears again (e.g. for *o*-terphenyl). Angell also compared the kinetics of mechanical relaxation with that of structural relaxation (enthalpy and volume) for different types of glasses and concluded that structural relaxation always proceeds more slowly than mechanical. On the whole, the temperature dependence of structural relaxation can be non-Arrhenius, although the viscosity may obey the Arrhenius law in the same temperature range. This is clear if we remember that according to Moynihan's theory, different structural parameters have different relaxation kinetics whereas the viscosity, being measured in the equilibrium state, is always defined by the slowest times of mechanical relaxation. Angell found in addition that as the structural relaxation process evolves the index β of the stretched exponential law for mechanical relaxation decreases and becomes stabilized. Thus there must not be any unique correlation between glass fragility (the viscosity characteristic of equilibrium state) and the exponent β for structural relaxation (approaching equilibrium). Nevertheless, the manifestation of fragility, as is often considered, can be interpreted in terms of higher or lower homogeneity of the spatial distribution of potential barriers. This may also lead to the evolution of a wide spectrum of structural relaxation times (see, e.g., Angell,[277] Stillinger[124], Tagliaferro[308]). Such an approach, however, is beyond the scope of the Adam-Gibbs theory, and its analysis is omitted here.

Theoretical aspects of the Adam-Gibbs model have been criticized by Goldstein.[181,309] In particular, it was pointed out that the real configurational contribution of the heat capacity and entropy of a glass at T_g can vary within wide limits, and other contributions can be comparable to them (this was discussed earlier). Also, the model of relaxation assuming a configuration change by the variation of the position of structure beads leaves undefined the cause of mass transfer with viscous flow. Such a model corresponds more closely to delayed elastic reversible deformation. For such relaxation, local stresses spreading in the same direction, when macroscopic stresses have been removed from the sample, should have resulted in a recurrent change of configuration because the relaxation causes microstresses of the opposite sign in other directions. The theory has merit in the connection of the viscous flow mechanism (and of the value of T_g) with the enthalpy (the entropy), which is nearer to a baric dependence of T_g, as mentioned in the previous section. We do not discuss the problem of the baric dependence of relaxational phenomena in terms of the Adam-Gibbs model because thus far there is no uniquely determined correlation with experiment (see, e.g., Richet[302] and Gupta[310]).

E. VALENCE-CONFIGURATIONAL THEORY

This theory is a version of activation theory that considers the existence of interatomic or intermolecular bonds, the cooperativity of atomic displacements, and configurational changes in liquid associated with temperature. Being a kind of molecular-kinetic theory, it defines flow processes in terms of a rheological model. Some of its parameters are connected with the instantaneous elastic characteristics of melt. Unlike the theories described above, this one contains only one arbitrarily selected parameter for the temperature dependence, which is the same for different classes of materials. The main goal of the theory is to explain quantitatively the connection between "glass length" ("long" and "short" or "strong" and "fragile" glasses) and the chemical structure of glassforming structural units. The basic concepts of the theory and its experimental verification are presented in papers by Nemilov.[311,312] In a survey by the same author,[275] the reader will find a more precise general discussion of the concepts. There is also a list of references to numerous papers appearing since 1964 in which the theory was used to analyze glass structure based on investigations of viscosity and elastic properties.

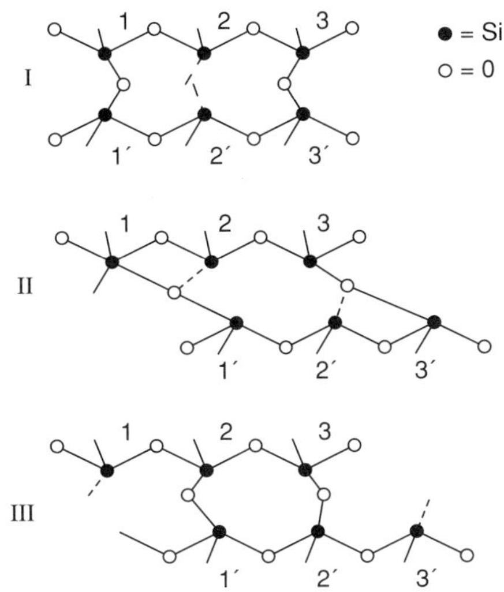

FIGURE 34. Bond-switching scheme according to the "bridge-vacancy" mechanism. There is no oxygen atom between silicon atoms 2 and 2′. The double dashes stand for the unsaturated valence of silicon. (I) the initial state, (II) the shift has resulted in straining the Si–0 bonds and in bringing together an oxygen atom and an unsaturated silicon valence, (III) the final state after bond switching: layers (1, 2, 3, …) and (1′, 2′, 3′, …) have undergone a relative shift for a distance Si–0–Si. The oxygen atoms have changed partners (1 and 1′ for 2 and 1′, etc.). (From Nemilov, S.V., *Sov. J. Glass Phys. Chem.* (Engl. transl. by Interperiodica Publ.), 18, 1, 1992. With permission.)

1. Bond Switching

The theory of chemical kinetics treats viscous flow as an activated movement of atoms (involving the concept of a free activation energy barrier) in which the bridging atoms (structural links, beads) change their hosts (e.g., silicon atoms) with the participation of vacancies (e.g., broken Si–O bonds), denoted by a superscript*,

$$1,2+3^* \rightarrow 1^*+2,3 \tag{324}$$

by a bond switching mechanism. The type of thermodynamic equilibrium structural defect is not specified in the theory because the bond switching energies are not calculated. Such defects may be oxygen vacancies, nonbridging oxygen atoms, etc. A scheme presented in Figure 34 produces a picture of the spatial character of the process. For large amplitude atomic vibrations due to anharmonicity (or to "soft" modes), atomic displacements occur that make possible the formation of an intermediate polynuclear complex. At the local stress stimulating the process, the complex dissociates and a local shear occurs. Until switching occurs, the deformation persists reversibly and corresponds to a delayed-elastic deformation. Similar schemes were suggested in the 1940s and 50s by Douglas,[313] Myuller,[314,315] and Smyth et al.,[316] and subsequently by many other researchers.

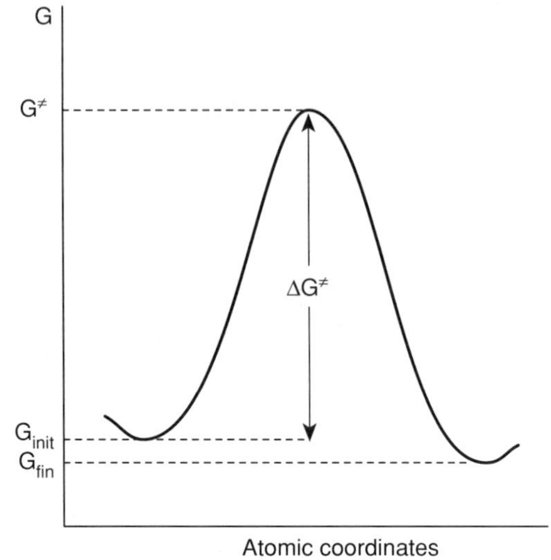

FIGURE 35. Free energy changes of activated-complex formation, $\Delta G^{\ddagger} = G^{\ddagger} - G_{init}$; $\Delta G_{reaction} = G_{fin} - G_{init}$.

Much data on isotope diffusion (oxygen, fluorine, etc.) in glass melts (see, for example, References 317 and 318) suggests that bridging atoms participate in viscous flow. The oxygen atom exchange between two neighboring tetrahedra is apparent in the results of 2D NMR chemical exchange spectroscopy (Farnan and Stebbins[319]). As in the theory of the transition state,[320] the free energy change at complex formation is represented by the scheme in Figure 35.

2. Free Energy of Activation and Activation Entropy

In the theory, the key parameters are the process rate constant (k) and the free activation energy ($\Delta G^{\ddagger}$),

$$k = k_0 \exp\left(-\Delta G^{\ddagger} / RT\right) \tag{325}$$

($R = k_B N$), the primary task being the modeling of k_0 (the value k_{0T}) and of the potential barrier $\Delta G^{\ddagger}$. The observed or apparent activation energies E in Figure 32 (E_1, E_2, E_3) are the activation heats

$$E \equiv \Delta H^{\ddagger} = \Delta G^{\ddagger} + T\Delta S^{\ddagger} \tag{326}$$

The activation entropies are defined as $\Delta G^{\ddagger}$ temperature coefficients,

$$\Delta S^{\ddagger} = -\partial\left(\Delta G^{\ddagger}\right) / \partial T \tag{327}$$

The discrepancy between the values of k_0 found by the theory (k_{0T}) and by experiment (k_{0E}) is connected with $\Delta S^{\ddagger}$ by the relation

$$k_{0E} = k_{0T} \exp\left(\Delta S^{\ddagger} / R\right) \tag{328}$$

The value of $\Delta G^{\ddagger}$ for the experimental k at temperature T is found simply from Equation (325),

$$\Delta G^{\ddagger} = RT \ln\left(k_{0T} / k\right) \tag{329}$$

The viscosity coefficient η is treated in a similar way,

$$\Delta G^{\ddagger} = RT \ln\left(\eta / \eta_{0T}\right) \tag{330}$$

An indication of significant activation entropies is the fact that the viscosity change is non-Arrhenius and/or a discrepancy between the theoretically calculated (η_{0T}) and experimentally defined (η_{0E}) values of the preexponential factor at a given temperature,

$$\Delta S^{\ddagger} = R \cdot \ln\left(\eta_{0T} / \eta_{0E}\right) \tag{331}$$

It should be noted that Eyring's theory of the transition state[320] is the most popular but is only one among many versions of the chemical kinetic theory. Interpretating significant activation entropies (over R) in terms of Eyring's theory as a measure of the difference between the transition complex entropy and that of the initial state is, however, impossible in principle.[321] It is noteworthy that in his last works Eyring never used these concepts; viscous flow peculiarities connected with the glass transition were reduced solely to the concept of free volume (see, for example, Reference 322).

From studies of dielectric relaxation in polymers[259,323] in the early 1940s and later biomolecule reaction kinetics,[324] it was concluded that such a model was inadequate. Significant activation entropies (of about 40 to 400 $j \cdot K^{-d}\, mol^{-1}$) were attributed to process cooperativity (Eley[324]) from which it appeared that the way of calculating k_{0T} had to be changed. In the Eyring theory,[320] for monomolecular process

$$k = \left(k_B T / h\right) \cdot \kappa \cdot \exp\left(-\Delta G^{\ddagger} / RT\right) \tag{332}$$

(κ is the transmission coefficient which is usually assumed to be 1), whereas at cooperative activation (see Reference 325)

$$k = \nu_0 \frac{\left(\Delta G^{\ddagger} / RT\right)^{s-1}}{(s-1)!} \exp\left(-\Delta G^{\ddagger} / RT\right) \tag{333}$$

Here s is the number of square terms into which kinetic energy equal to the activation energy $\Delta G^{\ddagger}/N = \Delta\varepsilon$ is resolved, and ν_0 is the mean thermal vibrational frequency (collisions, $\nu_0 \simeq k_B T/h$). A somewhat different meaning may be ascribed to s (see References 325 to 327), but all models have in common the fact that for the activation of small molecules in a gas or for a jump in a chain, s = 1, and Relation (333) reduces to (332). For activation of atomic displacements in a condensed state, including solids, where fluctuations arise from the superposition of vibrational modes in a three-dimensional medium, s = 2. (Reference 327) Relation (333) at s = 2 results if one has assumed large amplitude, anharmonic vibrations;[327] s = 2 corresponds to fluctuations observed when two weakly coupled harmonic oscillators exchange places.[328] The energy of oscillator interaction is assumed to be zero, but the activation energy is shared between the two oscillators. For s = 2, which applies to the hypothetical viscous flow in glasses mentioned above, we get

$$k = \left(\Delta G^{\ddagger} / Nh\right) \exp\left(-\Delta G^{\ddagger} / RT\right) \tag{334}$$

or, per particle,

$$k = (\Delta\varepsilon / h) \exp\left(-\Delta\varepsilon / k_B T\right) \tag{335}$$

This relation has been known in chemical kinetics since 1920 as Dushman's empirical equation.[329]

Relation (332) therefore presents a particular case. It follows from the activation conditions that are created in glass in the T_g region (see our discussion on structural changes in glasses above T_g) that to describe viscous flow, Relation (334) is most justified. The general relation (333) is derived strictly (Hinshelwood,[326] Polanyi and Wigner,[327] Moelwyn-Hughes[328]).

The particular case suggested by Eyring's theory for the interpretation of the process mechanism, however, does not assume that the very method of the activation values calculations from experiment by Formulas (325) to (331) becomes particular too. This had appeared in the theory of chemical kinetics even before the formulation of the transition state theory. The procedure is always valid on the condition that the activation energy is treated as a local change of free energy and can be interpreted broadly depending on the model used for the preexponential factor calculation.[330] Almost the same situation applied to molecular modeling of viscous flow.

3. The Volume of Activating Units

The expression for viscosity in Eyring's theory has the form

$$\eta = \left(Nh / V_{act}\right) \exp\left(\Delta G^{\ddagger} / RT\right) \tag{336}$$

where N is Avogadro's number, h is Planck's constant, and V_{act}/N is a molecular-like volume (i.e., the volume of the flow unit). Assuming the hole flow mechanism, this theory yields correlations between $\Delta G^{\ddagger}$ and the heat of molar vaporization for molecular liquids ($\Delta G \approx (1/4) \cdot \Delta H_{vap}$).

We have shown[331] that both in Myuller's bond switching theory[314,315] and in all of the other proceeding theories, although different assumptions are made regarding the process mechanism, the expressions for the preexponential factor differ from one proposed by Eyring only by a factor of the order of 1, which has the meaning of the packing density coefficient β,

$$\beta = V_{act} / \left(N \delta_1 \delta_2 \delta_3\right) \tag{337}$$

where δ_1, δ_2, and δ_3 are the linear dimensions of a kinetic unit. Thus, whatever the proposed mechanism of energy expenditure (bond switching, jumps of molecules to vacancies, etc.) the most general relation for the viscous flow coefficient is the Eyring equation (336). For molecular substances, V_{act} may correspond to a molar volume whereas for substances with a spatial distribution of bridging bonds, V_{act} may approximate the volume of a bridging atom.

Using methods of irreversible thermodynamics, Nemilov[275] showed that for viscous relaxation the rate constant coincided with the reciprocal Maxwell relaxation time τ_M,

$$k = 1/\tau_M \tag{338}$$

and the way of determining the stationary viscosity was equivalent to investigating stress relaxation at $t \rightarrow 0$. Thus the fractal correlation (214) has no connection with the problem of viscous flow.

Using the Maxwell relation

$$F_\infty / \eta = 1/\tau_M \tag{339}$$

where F_∞ is the instantaneous shear modulus, we can verify what equation for the relaxation time is better. When using Relation (332), one finds

$$F_\infty \cdot V_{act} = Nk_B T \tag{340}$$

The absurdity of such a situation was pointed out by Frenkel'.[332] If one uses Relation (334), then

$$F_\infty \cdot V_{act} = \Delta G^\ddagger \tag{341}$$

This relation, as precise as the coefficient about 1, contains the condition of fluctuational transformation of the elastic element of the Maxwell body into the viscous element (as was shown by means of local elasticity theory by Nemilov.[275]* Indeed, only those structural units that can overcome the potential

* The model for the viscoelastic Maxwell body contains one elastic and one viscous element that are subsequently connected. The complication of the model (combination of a number of different η_i and $F_{\infty i}$ elements) results in the appearance of delayed elasticity provided there is complete reversibility of the elastic deformation. Such sophistication does not change our result in principle.

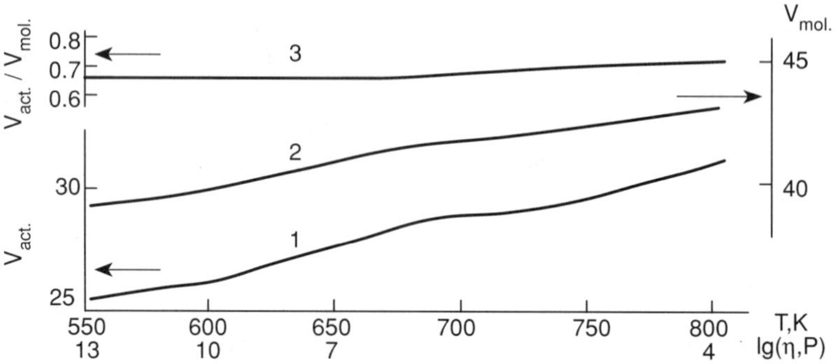

FIGURE 36. Temperature and viscosity dependencies of (1) V_{act}, (2) V_{mol} ($cm^3 \cdot g\text{-}mol^{-1}$), and (3) V_{act}/V_{mol} for a B_2O_3 melt. (From Nemilov, S. V., *Sov. J. Glass Phys. Chem.* (Engl. transl. by Interperodica Publ.), 18, 1, 1992. With permission.)

barrier of viscous flow kinetically can become viscous elements. At $T \rightarrow 0$ all elements become elastic, whereas at $T \rightarrow \infty$ all become viscous. $\Delta G^{\ddagger}$ defines the value of the constant for the thermodynamic equilibrium between elastic and viscous elements. Remember Mooney's work! [292]

Relation (341) in the form indicated was derived by Nemilov.[331,333] However, it should be noted that as long ago as 1943 Tobolsky et al.,[334] having presumed a specific cosine shape of the interatomic potential, obtained a similar correlation,

$$N\delta_1\delta_2\delta_3 \cdot F_\infty = 2\pi^2 \cdot \Delta G^{\ddagger} \tag{342}$$

In our treatment the existence of Relation (341) does not depend on the particular form of the potential and the coefficient $2\pi^2 = 19.7$ is absent.

Geometrical modeling of the bond switching process at significant deviations of bridging oxygen atoms from their equilibrium position led to the conclusion[312] that for three-dimensional structures like SiO_2, the maximum possible value of $\Delta G^{\ddagger}$ is of the order of a single bond breaking energy and V_{act} is approximately equal to the volume per oxygen atom. Planar networks formed by triangles of the $BO_{3/2}$ type must have a V_{act} of about twice the oxygen atom volume. The value of V_{act} is calculated from experiment at any temperature according to Equation (341) if the instantaneous shear modulus, the viscosity at temperature T, and $\lg\eta_0$ at $T \rightarrow \infty$ are known. V_{act} and V_{mol} for B_2O_3 are compared in Figure 36 in a wide temperature range. Obviously, $V_{act}/V_{mol} \simeq {}^2/_3$, exactly as expected. For other substances, the instantaneous moduli are known only for the region of the solid state of glass.

As the temperature decreases, the shear modulus, F_∞ progressively approaches the value measured at room temperature; below T_g it increases slightly (by 5–10%). Assuming that it would be roughly justified to identify F_∞ at T_g and F_∞ at 20 °C, we calculate V_{act} by Relation (341) for temperatures at viscosities of $1 \cdot 10^{16}$ P* from the shear moduli of solid glasses using the

* *Addition in Proof:* This value is close to lower temperature limit of optical glasses annealing region.

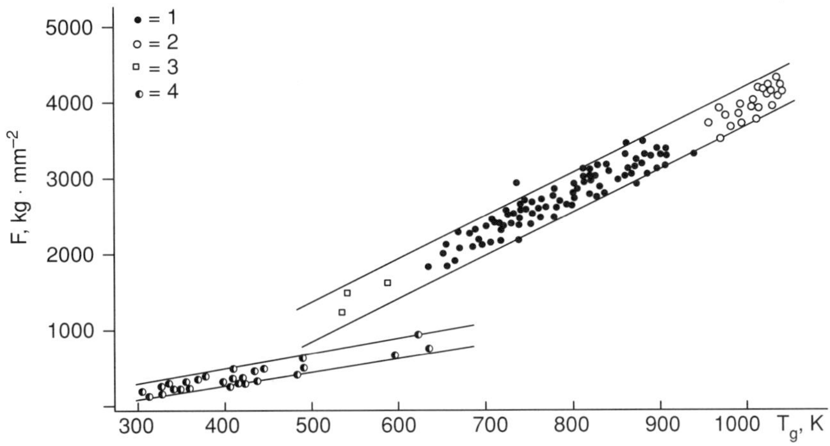

FIGURE 37. Shear modulus F of glasses as a function of T_g [for lg(η,P) = 13]: (1) silica-based optical glasses, (2) germanate glasses containing La_2O_3, Ga_2O_3, Ta_2O_5, etc., (3) BeF_2–based glasses, (4) chalcogenide glasses. (From Nemilov, S. V., *Sov. J. Glass Phys. Chem.* (Engl. transl. by Interperiodica Publ.), 18, 1, 1992. With permission.)

successive approximation method. For a large variety of oxide glasses (silicate, borate, metaphosphate, borosilicate, lanthania, germanate, and others), without exception V_{act} was found to range from 9 to 14 $cm^3 \cdot mol^{-1}$. On the whole, the values of V_{act} correlate with those of the physical volume per oxygen atom (calculated on the basis of density), hence the ratio of V_{act} to the volume of an oxygen atom V_{oxygen} remains about constant, $V_{act} \simeq (0.9 \div 1.1)V_{oxygen}$.[275] For chalcogenide glasses, V_{act} is approximately twice as large and corresponds to the volume per chalcogen atom in the As–S–As, As–Se–As, and other bridges calculated by the same method. Only meager data for metallic glasses are available. For one such glass ($Pd_{77.5}Cu_6Si_{16.5}$) both the viscosity in the glass transition region and the shear modulus are known, from which it was found that V_{act} = 6.3 $cm^3 \cdot$ g-at.$^{-1}$ (Reference 275). The mean gram-atom volume calculated on the basis of atomic radii is 6.0.

The correlation between F and T_g (= 10^{13}P) for different glass types derived from the validity of Relations (341) and (330) is illustrated in Figure 37. For chalcogenide glasses, the slope of this dependence is half as large as it is for oxide glasses because V_{act} is about twice as large. For BeF_2-based glasses the slope is the same as it is for oxide glasses because the fluorine atom volume is close to that of an oxygen atom.

Relation (341) explains why all correlations of elastic moduli changes with composition in multicomponent vitreous systems always appear to be approximately the same as those for the glass transition temperature changes with composition.

In most cases, the viscosity of low-molecular organic glasses has not been studied and only approximate values corresponding to the viscosity of 10^{13} P are known. Using the value of shear modulus for the vitreous state near T_g, one can estimate values of V_{act}, which, for these materials, turn out to be considerably

Table 4
Comparison of the V_{act} values calculated by Relations (330) and (341), and the molar volume for organic glasses

Substance	V_{mol}, $cm^3 \cdot mol^{-1}$	T_g, K	$F_\infty \cdot 10^{-9}$, Pa, by T_g	V_{act}, $cm^3 \cdot mol^{-1}$	V_{mol}/V_{act}	Number of functional groups
glycerol	70	183	3.5	17	4.1	3 (OH)
methylcyclohexane	115	85	1.62	16.6	6.9	7 (CH_x)
butyl chloride	81.4	98	2.1	14.7	5.5	5 (CH_x)
methylcyclopentane	91.6	78	1.96	12.6	7.3	6 (CH_x)
2-methylpentane	103.2	80	1.68	15.0	6.9	6 (CH_x)

From Nemilov, S. V., *Sov. J. Glass Phys. Chem.* (Engl. translation by Interperiodica Publ.), 18, 1, 1992. With permission.

smaller than the molar volume. For example, for glycerol at T_g, V_{act} . 17 and V_{mol} = 70 $cm^3 \cdot mol^{-1}$. In this substance the intermolecular interaction is governed predominantly by hydrogen bonds. The number of OH groups in a molecule (of which there are three) seems to be the factor that determines the ratio V_{mol}/V_{act}. For other substances without hydrogen bonds, the ratio V_{mol}/V_{act} is likewise the same, with an uncertainty of $< \pm 1$, as the number of functional CH_x groups responsible for van der Waals interaction (Table 4). Here viscous flow probably proceeds by a segmentary mechanism. In the case of an activated viscous flow, only fragments rather than whole molecules are displaced with the disruption of intermolecular bonds, such as occurs for the flow in polymers.[335]

Nemilov[275] has shown that the conclusion about the atomic mechanism of viscous flow in inorganic glasses and about the segmentary mechanism for organic substances conforms with the results of the investigation of non-Newtonian flow (at large shear stresses).

The above data illustrate that for calculating the theoretical value of the preexponential factor η_{OT} in Formula (336) it is necessary to use the values of V_{act} that yield the value $\lg\eta_{OT} \approx -3.5$ when the viscosity is expressed in Poises. In this case the typical temperature dependencies of activation enthalpy, free activation energy, and activation entropy are of the form presented in Figure 38. There is no need to point out the obvious correlation between the characteristic region of these parameter changes and the characteristic region of viscosity changes (see Figure 32). Also, it is obvious that the small values of the activation entropy correspond to "long" or "strong" glasses whereas the significant ones for "short" or "fragile" glasses are in the range from 10^{13} to 10^9 P.

4. Free Energy of Activation and Particle Interaction Energy

The free energy of activation for equilibrium melts attains its maximum values at T_g and at lower temperatures either remains constant or only increases

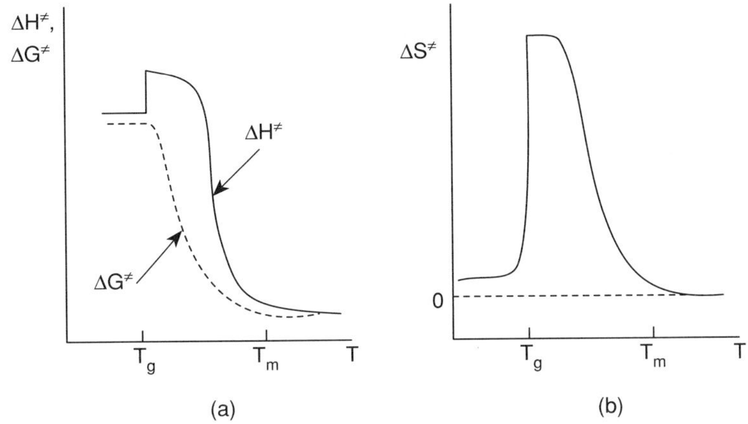

FIGURE 38. General pattern of changes in $\Delta G^{\ddagger}$ and $\Delta H^{\ddagger}$ (a) and $\Delta S^{\ddagger}$ (b) of glassforming melts and glasses with temperature. (From Nemilov, S. V., *Sov. J. Glass Phys. Chem.* (Engl. transl. by Interperiodica Publ.), 18, 1, 1992. With permission.)

very slightly. The apparent activation energy ($\Delta H^{\ddagger}$) cannot be identified with the energy of interatomic interaction because it includes the entropy term $T\Delta S^{\ddagger}$. For this reason, the $\Delta H^{\ddagger}$ of low-melting-point glasses, such as selenium ($T_g \sim 300$ K), is 1670 kJ · mol^{-1} whereas that of high-temperature SiO_2 ($T_g \sim 1450$ K) is only 525 kJ · mol^{-1}. Obviously, it is meaningless to identify the value of $\Delta H^{\ddagger}$ with any characteristic of thc intcraction energy. On the other hand, the free activation energy near T_g, when it increases, must approach the bond dissociation energy (Myuller[314,315]). For glasses with a spatially continuous bond network at viscosities of ~10^{16} P, these energies ($\Delta G^{\ddagger}$ and E_d, kJ · mol^{-1}) are indeed comparable in a rough approximation: SiO_2 (452 and 462), GeO_2 (305 and 352), BeF_2(230 and 377), As_2Se_3(176 and 224), and $GeSe_2$ (190 and 225).[336] The same values of $\Delta G^{\ddagger}$ can be calculated from the shear modulus and the volume of a bridging atom by Relation (341) and the glass transition temperatures from the bond strength, E_d, and volume.[337] Such cases are, however, relatively rare. First, the spatial bond network of a glass may consist of almost planar pyramids of the $BO_{3/2}$ type that make up layers or of chains and rings as in the case of selenium. Then local displacements and shifts can in principle occur without appreciably affecting the strong covalent bonds. Second, the modifier (i.e., alkali and alkaline earth) ions break up the spatial network of covalent bonds, thus leading to the superposition of ionic (Coulomb) interactions; this indirectly affects the bridging bond switching energy by reducing this value in comparison with the bond dissociation energy. This is clear from the fact that when the total interaction energies for covalent and ionic bonds are similar, the elastic constants for the former are always much higher than for the latter. Therefore, at the same atomic displacements (Δr) required for switching, the local change in the free energy is always greater for substances with a high degree of bond covalency than for those with more ionic interactions (Myuller,[338] Figure 39). For many glasses that do not contain alkali

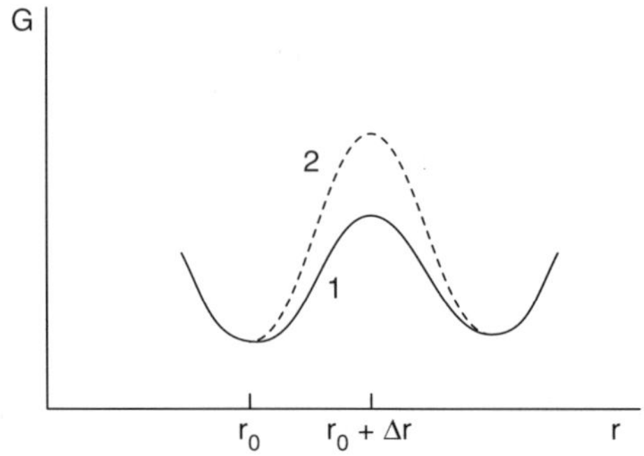

FIGURE 39. Free energy change at switching for more ionic (1) and more covalent (2) bonds of equal energies.

cations (e.g., glasses based on GeO_2, La_2O_3, Ta_2O_5, Y_2O_3, etc.), the free activation energies are extremely high (330–380 kJ · mol^{-1}) and come very close to the energies of single metal-oxygen bonds. However, a more precise comparison of $\Delta G^{\ddagger}$ and E_d in this case is hampered because there is no way to calculate reliably E_d due to the lack of knowledge of the real coordination numbers in glasses.

Glass transition temperatures and, accordingly, $\Delta G^{\ddagger}$ at T_g always increase with increasing atomization energy of the glass network. The reader will find a more detailed discussion of the ramifications and correlations in Reference 275.

It should be noted, however, that bridging-bond dissociation energies (if they are calculated as atomization energies) imply the dissociation of the condensed system into atoms completely removed from one another. This kind of situation is not encountered in the microscopic processes of atomic displacements. By virtue of the short range of covalent forces and the high values of the bond elastic constants, the activation energies for viscous flow can only approach the bond dissociation energy.

Table 5 lists the values of single bond strengths in inorganic glasses calculated by the author as the atomization energies on the basis of the heats of formation of the compounds ($\Delta_f H^{\circ}_{298}$) and the atomization heats of the elements (ΔH°_{f298}) using the well-known relation

$$E_d = \left[-\Delta_f H^{\circ}_{298}\left(A_m X_n\right) + m \cdot \Delta H^{\circ}_{f298}(A) + n \cdot \Delta H^{\circ}_{f298}(X)\right] / mK \qquad (343)$$

where K is the number of atoms of X species surrounding one A atom in the compound formula $A_m X_n$. Only references to the most recent data on $\Delta_f H^{\circ}_{298}$ of the compounds are listed in Table 5. Remaining thermodynamic values are obtained from reference books.[343–345] The values of E_d may prove helpful for estimating the effects of different glass components on T_g.

Glasses containing ionogenic components have Si–O bond energies, according X-ray emission spectroscopy data, approximately 20% lower than in SiO_2 (Dodd and Glen[346]), but the thermodynamic calculations do not confirm this.[339] For the latter, however, it is necessary to take into account the

Table 5
Energies of the chemical bonds in glasses

Compound, A_mX_n	Coordination Number of A Atom	Bond	Bond energy, E_d, kJ · mol^{-1}	Additional References
SiO_2	4	Si–O	462	
GeO_2	4	Ge–O	352	
	6	Ge–O	235	
B_2O_3	3	B–O	520	
	4	B–O	390	
P_2O_5	4	P–O	354	339
		P=O	620	339
Al_2O_3	4	Al–O	383	
	6	Al–O	255	
TiO_2	6	Ti–O	319	
ZrO_2	6	Zr–O	366	
	8	Zr–O	274	
La_2O_3	6	La–O	284	
PbO	4	Pb–O	166	
	6	Pb–O	110	
	8	Pb–O	83	
BeF_2	4	Be–F	377	
ZrF_4	6	Zr–F	225	
	8	Zr–F	168	
$ZnCl_2$	4	Zn–Cl	197	
S	2	S–S	277	
As_2S_3	3	As–S	251	340
Se	2	Se–Se	223	
As_2Se_3	3	As–Se	224	341
GeS_2	4	Ge–S	269	
$GeSe_2$	4	Ge–Se	225	342

Mean error within ±2%.

strength of Si–O$^-$ nonbridging bonds and the strength of alkali ion bonds with oxygen ions at the same time; we do not discuss obvious difficulties of this kind. In silicates the degree of ionicity for all the Si–O bonds also changes (Pauling[347]).

Valuable information can be gleaned from an analysis of the dependencies of T_g on the chemical composition and the structure of low-molecular glasses (hydrocarbons, alcohols, etc.) and also that of glassy crystals.[160] As has been found, T_g can be calculated by additivity formulas,

$$T_g(K) = n_{CH_x} \cdot \Delta T_{g_{CH_x}} + n_{OH} \cdot \Delta T_{g_{OH}} + \cdots \tag{344}$$

where n_{CH_x} is the number of structural CH_3, CH_2, and CH groups, n_{OH} is the number of OH groups in a molecule, etc., and $\Delta T_{g_{CH_x}}$ and $\Delta T_{g_{OH}}$ are the partial contributions from these groups to T_g: $\Delta T_{g_{CH_x}} = 12.4 \pm 0.3$ K (irrespective of x), $\Delta T_{g_{OH}} = 63.1 \pm 1.1$ K, $\Delta T_{g_{COOH}} = 116$ K, and $\Delta T_{g_{SH}} = 39.4 \pm 1.4$ K. The data treatment by Lesikar[348] for normal alcohols brings to close values of $\Delta T_{g_{CH_x}}$ and $\Delta T_{g_{OH}}$ (9.9 and 71.3 K). Proceeding from the relation for relaxation time

$$\tau_{rel} = \tau_0 \exp\left(\Delta G^{\ddagger} / RT\right) \tag{345}$$

one finds that the additivity of T_g in terms of contributions from structural groups is equivalent to the additivity of the free energy of activation in terms of contributions from the same groups ($\ln\tau_0$ is about the same and $\ln\tau_{rel}$ at T_g is exactly the same for different glasses, ~ –27.6 and 4.6, respectively),

$$T_g = \sum_i n_i \Delta T_{g_i} = \frac{\sum_i n_i \Delta G_i^{\ddagger}}{R\left[\ln \tau_{rel}\left(T_g\right) - \ln \tau_0\right]} \tag{346}$$

Hence the softening (glass transition) of molecular substances is most probably a complex process that calls for all the structural groups of a molecule to be excited, although not all of them may take part in an elementary shift event simultaneously. This statement implies that, for randomly oriented molecules, relaxation throughout the entire volume requires that all structural groups be mobile. Otherwise, stresses would accumulate and the process would stop. This problem, however, needs further study.

Comparing the partial contributions $\Delta G_i^{\ddagger}$ with the interatomic interaction energies $\Delta G_i^{\ddagger}$ (van der Waals and various hydrogen bonds) and the potential barrier U_0 of torsional rotations of the respective groups in the gaseous molecules, we see that $\Delta H_i \simeq \Delta H_i \gtrless U_0$ (Table 6). The interatomic interactions energy seems to be the leading factor in determining the possibility of relaxation at T_g because $\Delta G_i^{\ddagger} \simeq \Delta H_i$. The free rotation of the groups is hindered in the case of van der Waals interactions (although this is not necessary to violate this type of interaction). In the case of hydrogen bonds, free rotation at T_g must occur easily because the disruption of such directed bonds and activation of relaxation require greater energy expenditure. Actual coincidence of the values of T_g for glasses and crystals with the properties of glass (glassy crystals) of the same composition (see Table 1) is rather noteworthy.* These states essentially differ by their configurational entropy at absolute zero, and at T_g the contributions to the entropy from the configurational and "phonon" components are different. Apparently only the similarity of their potential barriers may be the explanation for the similarity of their T_g values.

For polymers, the correlation between $\Delta G^{\ddagger}$ and the energy of a monomer unit interaction with its environment is the same, in principle.[275] However, the calculation of T_g for polymers is not always successful because the linearity of

* For cyclohexene (a glass of normal origin) $T_g = 78$ K.[171] Compare this with T_g for glassy crystals in Table 1.

Table 6
Comparison of the intermolecular potential energy (ΔH_i) in the liquid and solid states, potential barrier of free rotation in vapor (U_0), and partial contributions to the relaxation activation on energy ($\Delta G_i^\ddagger$) for different functional groups

Group	Interaction type	ΔH_i	$\Delta G_i^\ddagger$	U_0
		kJ · mol⁻¹		
CH_2	van der Waals, in liquid hydrocarbons and polymers	2.8 ± 0.4	3.3 ± 0.3	13.8 ± 1.2
OH	H-bonds in liquid alcohols	18.8 ± 2.1	16.9 ± 1.3	4.6 ± 0.3
SH	H-bonds in liquid thioles and solid hydrogen sulphide	12.3 ± 2.9	10.5 ± 0.8	2.9 ± 0.2
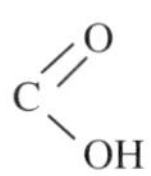	H-bonds in carbonic acids	29 ± 2	31 ± 2	2.0 ± 0.1

From Nemilov, S. V., *Fiz. i Khim. Stekla (Sov. J. Glass Phys. Chem.)*, 3, 423, 1977. With permission.

the relation between T_g and the cohesion energy is distorted by the chain stiffness factor (for more details, see Donth[349]).

5. Correlation Between Activation Entropy and Configurational Entropy

Before proceeding to the theory of the dependence of viscosity on temperature, two more empirical correlations should be mentioned. First, the increment of molar configurational entropy in the range from T_g to T_m, calculated by the formula

$$\Delta S_c = \int_{T_g}^{T_m} \frac{\Delta C_p}{T} dT \qquad (347)$$

satisfies a correlation, similar to that for T_g,

$$\Delta S_c = \sum_i n_i \cdot \Delta S_{c_i} = n_{CH_x} \cdot \Delta S_{c_{CH_x}} + n_{OH} \cdot \Delta S_{c_{OH}} + \cdots \qquad (348)$$

where the partial contributions $\Delta S_{c_{CH_x}}$, $\Delta S_{c_{OH}}$, $\Delta S_{c_{COOH}}$, and $\Delta S_{c_{SH}}$ are, respectively, 4.0 ± 0.25; 7.0 ± 0.5; 15.7 ± 0.6; and $11.0 \pm 1.0\ J \cdot K^{-1} mol^{-1}$ (Nemilov[160]). Remember that for all molecular organic glasses the ξ-values are nearly the same (see Table 1). Second, there is a direct proportionality between the activation entropy of viscous flow near T_g and the molar increment ΔS_c in the range T_g–T_m,

$$\Delta S^{\ddagger} = Z^{*}\left(\Delta S_c / \wp\right) \tag{349}$$

where $\wp$ is the number of the structure-linking beads per conventional molecule (Nemilov[350]). Figure 40 illustrates this correlation. The value of Z^* is ~92.5. $\wp$ is defined for inorganic substances as the number of bridging bonds and can be found by X-ray diffraction methods or by simple calculations from composition. For organic molecular substances, ΔS_c being calculated by Formula (347), the number $\wp$ corresponds to the total number of structural links. For *o*-terphenyl $\wp = 3$, because a molecule spatial structure is determined by three phenyl groups. At T_m, $\Delta S^{\ddagger}$ is much lower than that at T_g. This is because a liquids at T_m have a structural state such that the structural changes required for elementary bond switching become very small. The temperature dependencies of viscosity for glasses of most different types, at least in the T_g region, are defined by those changes of the configurational entropy by which the states of liquid at T_g and T_m differ (or at other temperature when $\Delta S^{\ddagger}$ vanishes).

Everything stated thus far still does not represent a theory of the temperature dependence of viscosity but only a set of theoretical facts and premises (correlations) necessary for solving the problem.

6. Theory of the Temperature Dependence of Viscosity

Thus far we have not taken into account the cause of the possible temperature dependence of $\Delta G^{\ddagger}$, which must define the value of $\Delta S^{\ddagger}$. The possibility of viscous relaxation is eventually reduced to the probability of fluctuational transformation of an elastic element (an atom vibrating near its equilibrium state) into a viscous element (an atom having the activation energy of viscous flow, $\Delta G^{\ddagger}$). The cooperativity has been considered only partly, as a way of the partitioning the activation energy among the degrees of freedom of an atom necessary to overcome the barrier. This treatment has not considered the environment of that atom and the possibility that the surrounding atoms could facilitate the switching of an individually selected bond to a new partner. Essentially, this is a matter of how the environment would respond to a sudden emergency of an atom with an excess energy $\Delta G^{\ddagger}$.

The term "cooperativity" implies its classic meaning, i.e. coordinated behavior of many particles in some finite volume that makes an elementary jump possible.

Angell[75,351] in his well-known "Bond Lattice Model," has mentioned that the activation energy must be the sum of a term independent on temperature and another term depending on it,

$$\Delta E = \Delta E_0 + \Delta E_1(T) \tag{350}$$

The first term in Equation (350) corresponds to an elementary defect state; the second one corresponds to the configurational changes connected with the

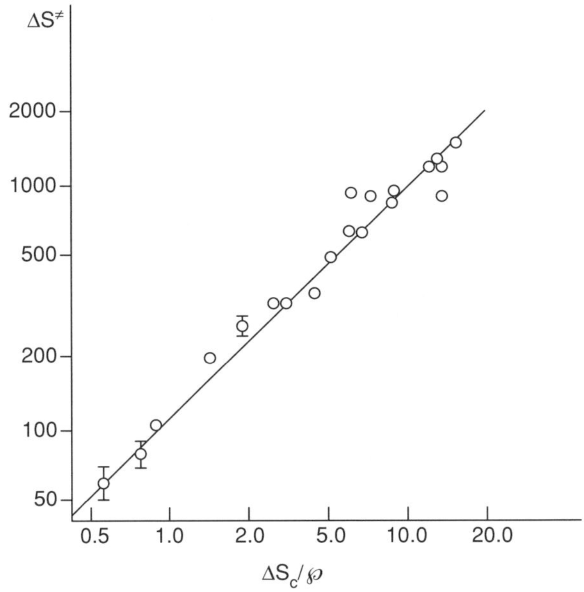

FIGURE 40. Viscous flow activation entropy $\Delta S^{\ddagger}$ in the T_g range (at 10^{13} P) as a function of the increment in the configurational entropy component for liquid (in the range from T_g to T_m) referred to the number of structural links ℘ per molecule.

The entropy values are given in $J \cdot mol^{-1} \cdot K^{-1}$ and plotted in log scales. The points (reading from left to right) correspond to GeO_2, SiO_2, BeF_2, Pyrex, soda-lime-silica window glass, $Na_2O \cdot 2SiO_2$, B_2O_3, As_2S_3, As_2Se_3, $BaO \cdot 4B_2O_3$, $SrO \cdot 2B_2O_3$, $Na_2O \cdot 4B_2O_3$, $PbO \cdot 2B_2O_3$, $BaO \cdot 2B_2O_3$, ($Na_2O \cdot 2B_2O_3$, $CaO \cdot 2B_2O_3$), (Se, glycerol, α-D-glucose, and *o*-terphenyl).

Source: Nemilov, S. V., *Fiz. i Khim. Stekla (Sov. J. Glass Phys. Chem.),* 2, 193 (1976). With permission.

lattice excited states. The second member bears the cooperative character of activation. The theory uses a two-parameter description of the configurational heat capacity by means of the Schottky function. Such an approach in this case seems correct because the theory does not have as its goal the description of the glass transition phenomenon. Those excited states identified with the fraction of atoms having excess heat capacity are treated by the theory formalism as analogous to the free volume; this eventually results in an equation of the FVT type. The relation analogous to (350) and having the same meaning was proposed by Filipovich and Kalinina[352] not long before this work.[351]

Our theory, as well as those mentioned, regards the probability of a bond switching in an equilibrium melt as the probability of a complex event. It is defined by the superposition of the probabilities of two events: bond switching and fluctuational change of the atomic environment. The latter fluctuations provide the decrease of potential barriers to their minimum levels. For this reason, rapid fluctuation in the energy appearing as thermal oscillations must occur within the limits of long-living fluctuations connected with small changes in the mutual displacements of atoms. This type of change is connected immediately with switchings and is defined by the vibrational and configurational states of equilibrium liquid, i.e., by its structure. If these fluctuations cannot proceed because of the system's nonergodicity, then bond switching takes place under unfavorable (as compared to an equilibrium ensemble) conditions, and the free activation energy changes only slightly with temperature, according to Relation (336). Thus, the free activation energy appears to be connected with these temperature and viscosity at which the structure fixation at the glass transition has occurred (see Figure 41). Denoting by $\Delta\mu_0$ the minimum bond switching potential in an equilibrium structure, by $\Delta\mu_c$ the configurational order transformation potential around the switching site per structural link, and by Z* the minimum number of structural links (beads) involved in such a transformation, we get for $\Delta\mu = \Delta G^{\ddagger}/N$ the expression

$$\Delta\mu = \Delta\mu_0 + Z^* \cdot \Delta\mu_c \tag{351}$$

In the general case $\Delta\mu_0$ undoubtedly includes energy expenditure for defect formation. Assuming $\Delta\mu_0$ is independent on temperature, we have for $\Delta s = \Delta S/N$ the expression

$$\Delta s = -Z^* \left(\partial\mu_c / \partial T\right)_p \tag{352}$$

provided Z* is also constant. At variable Z* we have

$$\Delta S^{\ddagger} / N = -Z^* \left(\partial\Delta\mu_c / \partial T\right)_p - \Delta\mu_c \left(\partial Z^* / \partial T\right)_p \tag{353}$$

For $\Delta\mu_c$ calculation, i.e., for determining the local change of free energy connected with the minimum local change in the configurational entropy, that state (temperature and pressure) is taken as a reference for which $\Delta\mu_c = 0$, that is, when the level of configurational excitations is fairly high and the viscosity is characterized by $\Delta S^{\ddagger} = 0$. From this it is obvious that this temperature (T*) lies somewhere in the region of the melting point. Exact knowledge of it is not important. Then, at any temperature above T_g, the free activation energy is calculated by the formula

$$\Delta\mu = \Delta\mu_0 + Z^* \int_T^{T^*} dT' \int_T^{T'} \Delta c_p \, d\ln T'' \tag{354}$$

and the activation entropy is calculated by the formula

$$\Delta s = -Z^* \int_T^{T^*} \Delta c_p \, d\ln T \tag{355}$$

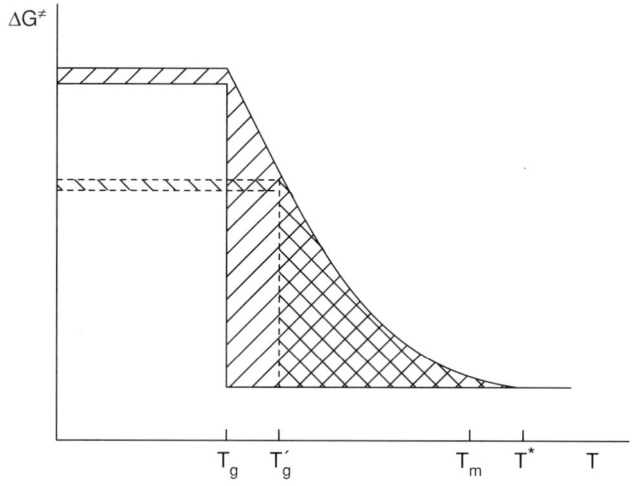

FIGURE 41. Theoretical relative magnitudes of $\Delta G^{\ddagger}$-components for various T_g. The bond-switching potential is represented by the lower boundary of the hatched areas. The cooperative order rearrangement potential is represented by the vertical range of the hatched areas. At $T < T_g$ practically all values of $\Delta G^{\ddagger}$ are determined by the product $F_\infty \cdot V_{act}$. At $T = T_g$ product $F_\infty \cdot V_{act} \simeq \Delta G^{\ddagger}$ corresponds to an instantaneous fluctuation in an elastic medium. The subsequent isothermal rearrangement of the environment reduces the value of the bond-switching potential. (From Nemilov, S. V., *Sov. J. Glass Phys. Chem.* (Engl. transl. by Interperiodica Publ.), 18, 1, 1992. With permission.)

(the dependence of Z^* on T is ignored). Δc_p is the configurational heat capacity component per link. The relative contributions to Δc_p and Δs by pure configuration changes and by the vibrational spectrum changes connected with these configurations is also not important. In this theory, the treatment of the configurational entropy bears a meaning different from that in the Adam-Gibbs theory. In the latter theory the configurational entropy determines the state of a system, whereas in our theory it is the fluctuational change of the state (Figure 42). Obviously, our theory cannot be reduced to that of Adam-Gibbs, although the general interpretation of the temperature dependence of viscosity is similar: the higher the temperature coefficient of viscosity, the more significant the configurational changes of structure occurring in the melt. The higher these levels are, the higher the heat capacity of the melt configuration.

Having ignored the temperature dependence of the preexponential factor in the expression

$$\eta = \eta_{0T} \exp\left[\left(\Delta\mu_0 + Z^* \Delta\mu_c\right) / k_B T\right] \tag{356}$$

we have for the apparent activation energy from our theory, the expression

$$\frac{\Delta H^{\ddagger}}{N} = k_B \frac{\partial \ln \eta}{\partial (1/T)} = \Delta\mu_0 + Z^* \cdot \Delta\mu_c(T)\left[1 + \frac{\partial \ln \Delta\mu_c}{\partial \ln T}\right] \tag{357}$$

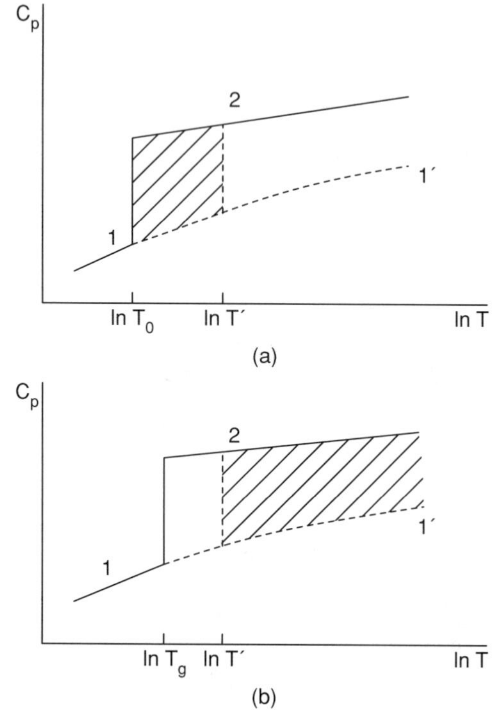

FIGURE 42. Calculations of entropy changes using the model proposed (a) by Adam and Gibbs and (b) by Nemilov. 1, 1′, heat capacity of glass and the phonon component of heat capacity of liquid; 2, heat capacity of liquid above T_0 (or T_g). (From Nemilov, S. V., *Sov. J. Glass Phys. Chem.* (Engl. transl. by Interperiodica Publ.), 18, 341, 1992. With permission.)

For the Adam-Gibbs theory, we have from Equation (320)

$$\frac{\Delta H^{\ddagger}}{N} = k_B \frac{\partial \ln \eta}{\partial (1/T)} = \frac{\Delta \mu \cdot s_c^*}{S_c}\left[1 + \frac{\partial \ln S_c}{\partial \ln T}\right] \tag{358}$$

In our theory, the main cause of significant $\Delta H^{\ddagger}$ values is the dependence of $\Delta\mu_c$ on temperature (the Z* vs. T dependence is not necessary), whereas in the Adam-Gibbs theory it is the dependence of S_c on temperature at constant $\Delta\mu$.

If the molar ΔC_p is known, as well as $\wp$ (the number of beads in a molecule), then for the known $\Delta H^{\ddagger}$ and $\lg\eta_{0E}$ for some temperature (T_{ref}), the calculation of $\lg\eta$ for lower temperatures proceeds from the formula

$$\lg \eta = \lg \eta_{0E} + \left\{ \Delta H^{\ddagger}_{T_{ref}} + \frac{Z^*}{\wp} \int_{T}^{T_{ref}} dT' \int_{T}^{T'} \Delta C_p \, d\ln T'' + \Delta S^{\ddagger}_{T_{ref}} \cdot \left(T_{ref} - T\right) \right\} / 2.303\,RT \tag{359}$$

The meaning of such a calculation is as follows. At T_{ref}, the values of $\Delta S^{\ddagger}_{T_{ref}}$ and $\Delta G^{\ddagger}_{T_{ref}}$ are calculated by Formulas (330) and (331), using the theoretical value of $\lg\eta_{OT} = \lg(Nh/V_{act}) \simeq -3.5,$). The viscosity and its temperature coefficient at T_{ref} need to be known. For lower temperatures, $\Delta G^{\ddagger}$ grows by the temperature coefficient ($\Delta S^{\ddagger}$) at T_{ref} having the experimental value of $\Delta S^{\ddagger}_{T_{ref}}$ and its increment at $T < T_{ref}$ is defined by the integral. The value of Z* is found by selecting the best possible agreement with the experimental viscosity. The viscosity is calculated as

$$\lg\eta = \lg\left(Nh / V_{act}\right) + \frac{\Delta G^{\ddagger}(T)}{2.303\,RT} \tag{360}$$

which is equivalent to the previous one, but where $\Delta G^{\ddagger}$ (T) has already been referred to a single structural bead and includes the factor Z* ($\Delta G^{\ddagger} = N \cdot \Delta\mu_0 + NZ^*\Delta\mu_c$). If at $T_{ref} = T_m$, $\Delta S^{\ddagger} \approx 0$, the term $\Delta S^{\ddagger}_{T_{ref}} \cdot (T_{ref} - T)$ in Formula (359) can be neglected.

7. Comparison of the Theory with Experiment

The temperature dependence of the form of Equations (359) and (360) does not assume a critical increase of viscosity with decreasing temperature as the FVT equation does. However, it also does not show a rigorous Arrhenius behavior in the high viscosity range. If ΔC_p is small enough or if ΔC_p increase with temperature, as in the case of SiO_2 or GeO_2, the theory gives a viscosity change law that is close to the Arrhenius one, and, within experimental error, conforms with experiment. At the same time, acknowledging that the minimum T_g value that might be attained by slow cooling cannot become lower than T_K (otherwise, admitting the existence of a second-order phase transition, at T_2), theoretical calculations predict a universal value of viscosity at that temperature for many glasses of $10^{20.6 \pm 0.7}$ P (Nemilov[312]). This value, however, is so high that such a temperature appears unattainable in the usual experiments for equilibrium viscosity.

We have calculated[311] the viscosity for many substances at $T_{ref} = T_m$. Throughout the whole viscosity range (from 10^{-2} – 10^4 P, depending on substance, to 10^{13} P) the viscosity may be calculated by Equation (359) at the values of Z* listed in Table 7. The values of $\wp$ were calculated based on the composition (the number of bridging bonds for inorganic glasses, the number of functional groups for organic molecules). In the case of $Na_2O \cdot 2B_2O_3$ we have used data from X-ray diffraction structure analysis. The comparison of calculations with experiment at constant Z* for this substance is given in Figure 43. For glycerol, α-D-glucose, and Se the values of Z* appeared to a significant extent to depend on temperature, although remaining of the same order of magnitude (~10^2). In the majority of cases, the small dependence of

Table 7
Comparison of the constant values of Z* that fit experiment to theory in the range from T_g to T_m, according to Nemilov.[311] For As_2S_3 see Reference 353.

Substance	℘	Z*	Substance	℘	Z*
SiO_2	2	92.5	$BaO \cdot 4B_2O_3$	13	92.5
GeO_2	2	92.5	$Na_2O \cdot 2B_2O_3$	6.5	92.5
BeF_2	2	92.5	$BaO \cdot 2B_2O_3$	7	92.5
B_2O_3	3	92.5	$SrO \cdot 2B_2O_3$	7	106
As_2S_3	3	55.5	$PbO \cdot 2B_2O_3$	7	87
	(5)[a]	92.5	Propanol	4	92.5
$Na_2O \cdot 2SiO_2$	3	92.5	Isobutanol	5	92.5
$Na_2O \cdot 4B_2O_3$	13	92.5	*o*-terphenyl	3	99
$K_2O \cdot 4B_2O_3$	13	92.5			

[a] In this substance the volumes of As and S atoms are nearly equal or comparable. Arsenic atoms may play the role of "beads" for bond-switching.

the calculated results on Z* at low viscosities does not show a reliable nonconstancy of Z*.

In Table 8 calculated viscosities are compared for (NBS USA) glass N 710 (mol.%, SiO_2 — 72.9; Na_2O — 8.8; CaO — 12.9; $(Sb_2O_3 + SO_3 + R_2O_3)$ — 0.3; $M = 61.9\ g \cdot mol^{-1}$). Scherer[306] has done the calculation according to the Adam-Gibbs theory using the equation

$$\eta = \eta_0 \exp\left[\frac{Q/RT}{\ln(T/T_2) + C(T - T_2)}\right] \tag{361}$$

and Q/R = 8052 K, $T_2 = 493.5$ K, $C = -8.0 \cdot 10^{-4}\ K^{-1}$, $\eta_0 = 1.13 \cdot 10^{-4}$ P. The heat capacity ($J \cdot K^{-1}g^{-1}$) has been approximated by the equation

$$\Delta c_p = 0.805 - 6.44 \cdot 10^{-4}\ T \tag{362}$$

which differs from the experimental relation[355]

$$\Delta c_p = 0.984 - 8.83 \cdot 10^{-4}\ T \tag{363}$$

Using the latter equation, the author, according to Relation (359), has calculated the viscosity for that glass assuming ℘ = 1.05 (the number of bridging bonds per mole of mass M) and Z* = 136.5. The agreement with the interpolated experimental data is equally good for both calculations.

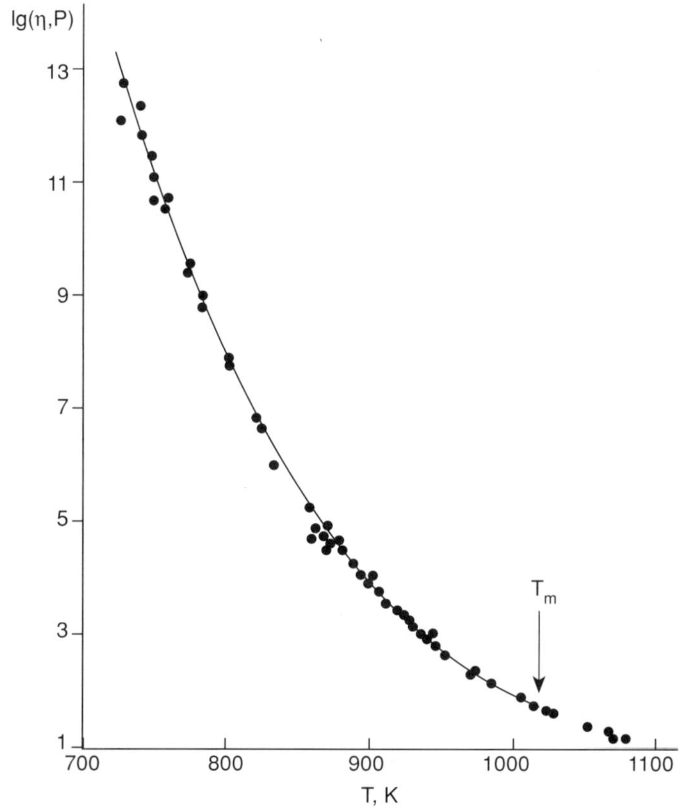

FIGURE 43. Viscosity of $Na_2O \cdot 2B_2O_3$ as calculated by the valence-configurational theory (solid line) compared with the experimental data obtained by different authors. (From Nemilov, S. V., *Fiz. i Khim. Stekla (Sov. J. Glass Phys. Chem.),* 4, 129, 1978. With permission.)

Figure 44 presents a comparison of experimental data for *o*-terphenyl with those calculated by our theory and by the Adam-Gibbs theory (Chang and Bestul[135]). Here we have used the value of $\Delta S^{\ddagger}$ at $T_{ref} = T_m = 329.35$ K (135.3 $J \cdot K^{-1} mol^{-1}$). At T_m, $\lg(\eta_{OE}, P) = -10.561$ and $\Delta H^{\ddagger} = 6.325$ kJ $\cdot$ mol^{-1}. This corresponds to $\Delta G^{\ddagger} = 18733$ J $\cdot$ mol^{-1} and $\lg(\eta,P) = -0.53$ at T_m. Instead of integrating the heat capacity for the calculation of ΔS_c, I have used the increments of the configurational entropy for the range of $T - T_m$ directly and

$$\Delta S_c = \left[S(l) - S(c)\right]_{T_m} - \left[S(l) - S(c)\right]_T \tag{364}$$

was calculated for the data in Reference 135. The value of $\lg(Nh/V_{act})$ is taken to be -3.5. The value of $Z^* = 101.6$ for which the theoretical curve has been calculated differs only slightly from the one (99) taken in Reference.[311] In the range of 10^{13} to 10^9 P the theoretical dependence essentially corresponds to the

Table 8
Comparison of viscosity values for the soda-lime-silica glass N 710 (NBS USA) calculated: by the Adam-Gibbs theory according to Scherer[306] [Equation (361)]; by Nemilov's theory [Equation (359)], and measured experimentally.[354]

Temperature, °C	lg(η,P) Scherer	Nemilov	Experiment[a]
520	14.84	14.81	14.84
540	13.71	13.71	13.71
560	12.72	12.73	12.72
580	11.84	11.86	11.84
600	11.07	11.08	11.07
620	10.37	10.38	10.37
640	9.75	9.75	9.75
660	9.19	9.19	9.19
680	8.68	8.674	8.674[b]

[a] Calculated by the equation $\lg(\eta,P) = -2.008 + 4669/[T - 242.9\ (°C)]$, which is recommended in Reference 354 for the best interpolation of the data obtained by the methods of fiber elongation and beam-bending (standard deviation within 0.06).

[b] The calculated data (Nemilov) for lower temperatures are referred, $T_{ref} = 680$ °C, to this viscosity value.

Arrhenius type and agrees within experimental error. The Adam-Gibbs theory does not yield such a result. The agreement with experiment is a bit worse at $\eta < 10^6$ P; this may be attributed to the small dependence of Z* on temperature. Complete agreement with the averaged experiment is obtained if in the range of 10^{13} to 10^7 P, Z* decreases only very slightly (from 102.4 to 96.3) and at lower viscosities (10^6– 10^2 P) this decrease becomes more significant (from 94 to 60). If the theory is correct, the explanation for this could be found in the overlapping of fluctuational regions in a low-viscous state.

In Reference 311 the reader would find other comparisons of theory with experiment and the prediction of the viscosity of ethanol up to 10^{13} P. A peculiarity of the theory is the universality of the parameter Z* value ($\sim 1 \cdot 10^2$) for various classes of substances. Ignoring the slight temperature changes in Z* required for complete agreement between the theory and experiment in the region of low viscosities, we find that for a constant Z* value the viscosity in the glass transition region is always well described and the values of $\Delta G^{\ddagger}$ and $\Delta S^{\ddagger}$ at T_g ($\eta = 10^{13}$P) correspond to that obtained experimentally.

In any version of the activation theory of viscosity, the range of the applicability of the concept of an activation barrier is confined to large viscosity values.[181] This is defined by the fact that the time of overcoming the potential

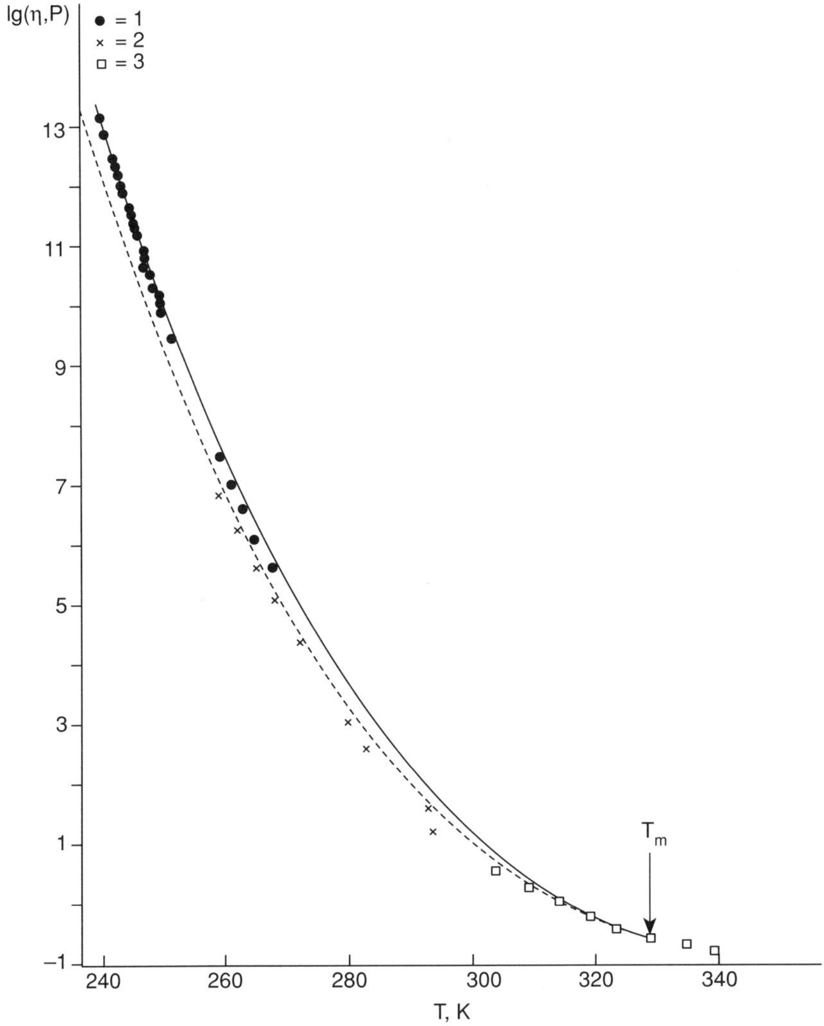

FIGURE 44. Viscosity for *o*-terphenyl as calculated by the valence-configurational theory (solid line) compared with that calculated by the Adam-Gibbs theory (dashed line, Chang and Bestul[135]). Experimental data: (1), Reference 276; (2), Reference 356; (3), Reference 357.

barrier must be less than the time required for the fluctuation to have disappeared. From estimates by Goldstein[181] and Moynihan[358], the smallest viscosities for which the activation theories apply are of the order of 10^1–10^2 P. This may also be the cause of the better agreement of our theory with experiment at higher viscosities.

In spite of the universal character of the parameter Z*, the linear dimension of the region of cooperative transformation of structure can vary significantly depending on interatomic distances or molecular sizes. In SiO_2 this region must include about 50 $SiO_{4/2}$ tetrahedra, whereas in cyclohexanol it must include about 14 molecules. Actually, this is the scale of medium-range structural order. With respect to structure, configurational changes at viscous flow are treated by this theory as only slight variations in mutual particle displacements. Bridge bending or folding, similar distortions of hydrogen bond orientation, and changes in the orientations of structural groups create the opportunity for spatial compensation of the local deformations that are necessary for altering the environment of an atom or a group and that overcome the switching barrier.

Many independent theoretical and experimental estimations (see references cited in References 275 and 311) give the same magnitude for the size of the cooperative rearrangement region. The theory of the temperature dependence of viscosity presented here, as well as in the Adam-Gibbs theory, reduces to the so-called "theory of configurational heat capacity." The valence structure of a substance superimposes its own, and thus far not profoundly understood, restrictions on its magnitude. One thing is obvious, namely that the character of spatial connectivity of network and the character of interatomic bonds for such theory are of great importance.

The question of the effect of pressure on viscosity is not considered here. Obviously the baric dependence of $\Delta G^{\ddagger}$ is defined by the baric dependencies of the associated components, and these factors are generally unknown. It is clear that V_{act}, which is involved in the expression $F_{\infty}V_{act} = \Delta G^{\ddagger}$, may not coincide with the value of $\Delta V^{\ddagger} = (\partial \Delta G^{\ddagger}/\partial p)$ because above T_g the configurational contribution into $\Delta G^{\ddagger}$ may be significant.[359]

8. Viscosity of Glasses with Frozen Structure

The viscosity of glasses with fixed structure is defined by the nonergodicity of the system below T_g, and no significant fluctuations of the configurational entropy are possible in it.[312] Therefore, below T_g the temperature dependence of the viscosity is determined by bond switching without significant reversible configurational rearrangements. The potential barrier $\Delta G^{\ddagger}$ almost corresponds to the bond switching energy, and the configurational component in the ideal case must be absent (Figure 41). Therefore, at $T < T_g$ it should be $\Delta S^{\ddagger} = 0$, and the viscosity below T_g should be calculated by the relation

$$\eta = \left(Nh / V_{act}\right)\exp\left[\frac{\Delta G^{\ddagger}\left(T_g\right)}{RT}\right] \tag{365}$$

Thus, for glass N 710, at $T_g = 793$ K and lgh = 14.83 at T_g, we have $DG^{\ddagger}(T_g) = 279$ $kJ \cdot mol^{-1}$, which coincides with the value obtained by Scherer[306] by the Adam-Gibbs theory (283) and is close to the experimental one (297). In principle the same value could be obtained without viscosity measurements by calculating $DG^{\ddagger}$ and lgh by the valence-configurational theory. Examples of such calculations for different glasses are given by Nemilov.[312] It is also simple to determine the parameter x in

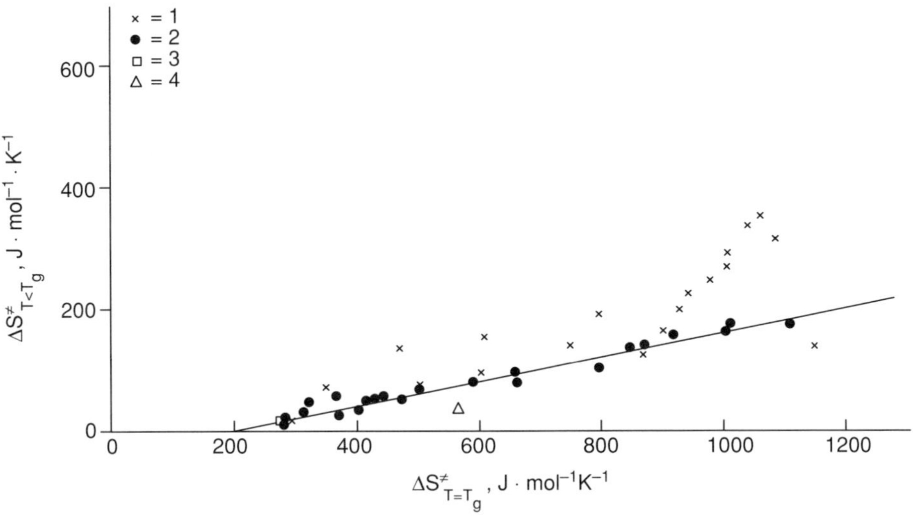

FIGURE 45. The dependence of $\Delta S^{\ddagger}_{T<T_g}$ (glasses with fixed structure) on $\Delta S^{\ddagger}_{T=T_g}$ (equilibrium melts at T_g). (1), data from Mazurin's laboratory[360] for simple oxide systems; (2), data from Nemilov's laboratory (measured by Sizov[361]) for optical glasses; (3), B_2O_3 (Reference 360); (4), NBS No. 710 (Reference 265).

Narayanaswamy's formula (221)

$$x^{-1} = \lim_{T \to T_g} \frac{\Delta H^{\ddagger}(T)}{\Delta G^{\ddagger}(T)} = \lim_{T \to T_g} \left(1 + \frac{T\Delta S^{\ddagger}}{\Delta G^{\ddagger}}\right) \tag{366}$$

which is necessary for engineering calculations of relaxation.

The accepted approximation $\Delta S^{\ddagger} = 0$ for glass below T_g is apparently too rigid. Numerous experiments show that as $\Delta S^{\ddagger}$ at T_g increases, the glass viscosity at $T < T_g$ is characterized by small but growing values of $\Delta S^{\ddagger}$ (Figure 45). At $\Delta S^{\ddagger}_{T \gtrsim T_g} < 200$, apparently $\Delta S^{\ddagger}_{T < T_g} \simeq 0$, as the theory assumes. For commercial optical glasses, at $\Delta S^{\ddagger}_{T \gtrsim T_g} > 200$, one may use the equation

$$\Delta S^{\ddagger}_{T<T_g} = 0.185\left(\Delta S^{\ddagger}_{T \gtrsim T_g} - 200\right), \mathrm{J \cdot mol^{-1} K^{-1}} \tag{367}$$

The cause of $\Delta S^{\ddagger}_{T < T_g} > 0$ is apparently in the fact that soft configurations representing localized states are preserved in the glass. These states do not result in equilibrium values of the configurational heat capacity (ΔC_p) by virtue of their localization and the nonergodicity of the system. However, the fraction of such states is rather large (0.1), and bond switching can take place in just such regions. The statistical description of the bond switching process in this case must differ from the one stated above. To associate the viscous flow below T_g with β-relaxation, this relaxation must apparently be connected with small activation entropies (Wu and Nagel[362]). Such an approach could also explain the small activation entropies for viscous flow.

Correlations can be established between the concepts of the free volume theory and those of the valence-configurational theory. Sanditov[283] has addressed this issue but we do not dwell upon it here.

Concluding Remarks

As envisioned in this book, as a melt is cooled below its melting point, metastable liquid gradually loses its excessive entropy compared to that of a crystal. This occurs due to the decrease of the population of the configurational part of the phase space, to the change of the phonon spectrum, and to other changes. Gradual structure ordering proceeds. Simultaneously the difference between the free energies of liquid and crystal increases, which, however, is regulated by the entropy relations. At T_g the difference between the free energy and the entropy of glass and internally equilibrated liquid appears and then increases. Below the glass transition temperature, translational displacements of atoms or molecules and disorder are frozen in. The system's approach to an energy minimum is attained by the peculiarities of its phonon spectrum differing from the crystal spectrum. Localized states with lower vibrational frequencies play a significant role here. Their appearance in glass is defined genetically by the structure of the liquid from which the glass has been obtained. These states play a significant part in the optimization of the correlations between the energy of the system and its entropy.

Strict correlations exist between the frozen energy and entropy values and those reversible changes of these functions that occur in liquid and glass. The reduced form of these correlations is universal for all glasses (the thermodynamic invariant of the vitreous state). This suggests the possibility of creating in principle a generalized theory of glass formation on the basis of statistical mechanics. The structure of glass and its frozen energy and entropy values are, however, specifically defined by both the individual character of each substance's "crystallochemical" structure and the kinetic regime of glass transition.

At the glass transition, the freezing of the structure parameters defines the different ways the entropy, volume, and shear stresses relaxation proceeds. This defines both the Prigogine-Defay Ratio being different from one and the specific kinetics of the reverse process of the glass state relaxing to the state of equilibrium liquid (Moynihan relation). The relaxation of properties having different thermodynamic meaning is defined by the kinetic correlations of the fractal type. These process parameters (the characteristic time and the index of stretched exponent) are close but still different. At the same time, in every case the activation energy appears close to that for shear relaxation, the latter being the most rapid process. The energy barriers in every case belong to the same distribution, and the elementary acts on the molecular level (e.g., bonds switching) are of the same type. However, the consecutive set of these acts ("process path") that finally results in structure formation and property relaxation must be individual for each property. Probably the "process path" of the structural relaxation component for each of the properties is as individual as the nonequilibrium thermodynamics structure parameter that describes this property change — and they are closely bound. It should be noted that the emergency of a fractal correlation is apparently always an indication of a condensed system approaching equilibrium. Any liquid above T_g may be driven from equilibrium by an appropriate influence (by any alternating field of appropriate frequency). Most problematic is the association between structural causes for the onset of fractal kinetic laws of structural relaxation and structural factors of non-Arrhenius behavior of kinetic characteristics (e.g., of viscosity) of metastable liquid at infinitesimal deviations from the equilibrium. This question is now being studied intensively.

When liquid is supercooled a significant viscosity rise occurs, resulting in a glass transition. All the available viscous flow theories explain this phenomenon, on the global scale, in the same way: "The system being cooled undergoes such changes in the components of its thermodynamic parameters (the free volume as part of the total volume or the configurational entropy as a part of the state entropy) which induce, like an avalanche, the viscosity increase in the system." This process mechanism may be interpreted in different ways. In particular, it could be assumed that lowering the temperature results in significantly increasing potential barriers and their magnitude differentiation connected with a concrete type of structural ordering (topological and structurally chemical*). If viscous flow is defined by the most rapid processes, the inhomogeneity within the set of potential barriers appears essential for a spectrum of property relaxation times (the stretched exponent index). We have not considered here the theories of the fractal coefficient β, for it would have caused us to deviate from the subject. During the past decade, tens of papers have been devoted to this problem and the number of original papers grows extremely rapidly (we point out only some of them: References 182, 245, 308, and 363 to 377).

All theories consider the cooperativity factor, which varies with temperature. It is cooperativity that defines all the peculiarities of the temperature dependence of viscosity and structural relaxation. In solid glass viscous flow processes are featured by low cooperativity and may be connected with β-relaxation and with localized states of soft modes. We have not dwelt on other approaches modeling the cooperativity. Worth mentioning are the models of the two-dimensional facilitated kinetic Ising model type which, although not glass transition models, are able to describe the spin flip process as a cooperative one and also give a theoretical expression for the stretched exponent index β (see, e.g., References 378 and 379). In the last decade a very interesting Mode Coupling Theory was suggested.[380,381] Thus far, however, it is far from thermodynamics. This theory successfully describes the critical development of spatial correlations of density fluctuations above T_g in good agreement with experiment.[382] The correlation of this theory with those considered here can be found in the survey by Nemilov.[261]

* Here we mean chemical equilibria displacement, fluctuational ordering as in supercritical phenomena of liquid–liquid immiscibility, pseudo-eutectic structure formation, etc.

This book considered only the liquid state as the genetic predecessor of the vitreous state. This limitation is not essential (Nemilov[383]). In the sol-gel method the vitreous state is achieved below T_g because a colloidal system, by the compositional change (through hydrolysis and its products' evaporation), transforms into a metastable liquid state. When amorphous films are deposited on a substrate, their initial state may not be comparable with a given state of liquid. By force of the thermodynamic Ostwald's Steps Rule (which has a rigorous thermodynamic basis[384]), such a state may emerge as an intermediate one before crystallization, and the amorphous film would transform into the state of glass, analogous to that obtained from liquid. Crystals' amorphization by pressure corresponds to the simplest case of the transition of a point from crystal area to a liquid one in a phase diagram (e.g., hexagonal ice,[385] α-quartz[386]). The structure of α-quartz that has been amorphized by pressure is practically the same as that of vitreous silica obtained from liquid.[387]

Electromagnetic field is a proper parameter of state of a system. Therefore, changes in the glass transition temperature and the viscosity under a photochemically active flux, in the thermodynamic sense, is quite natural.[388] Light causes similar structural changes as does temperature.[389] The well-known amorphization of quartz under a fast neutron flux is also a way of obtaining glass by changing a parameter of state of an open system.

The experimentally observed behavior of glass may always be understood and described in the language of thermodynamics.

In this book I have had to limit the subjects under consideration. I apologize to those authors whose works, because of this restriction, have not been mentioned here. Nevertheless, I do hope that this book may serve as an introduction to the problem and that it may prove helpful to those interested and active in vitreous state science.

References

1. Tammann, G. A., *Der Glaszustand,* L. Voss, Leipzig (1933).
2. Zachariasen, W. H., The atomic arrangement in glass, *J. Am. Chem. Soc.,* 54, 3841 (1932).
3. Frankenheim, M. L., *Die Lehre von der Cohäsion,* August Schulz und Komp., Breslau (1835), 389.
4. Tool, A. Q., Relation between inelastic deformability and thermal expansion of glass in its annealing range, *J. Am. Ceram. Soc.,* 29, 240 (1946).
5. Brawer, S., *Relaxation in Viscous Liquids and Glasses,* The American Ceramic Society Inc., Columbus, Ohio (1985).
6. Feltz, A., *Amorphe und Glasartige Anorganische Festkorper,* Jena Acad. Verlag, Berlin (1983).
7. Zarzycki, J., *Glasses and the Vitreous State,* Cambridge Univ. Press, Cambridge (1991).
8. Jones, G. O., *Glass,* Chapman & Hall, London (1971).
9. De Donder, Th., *L'Affinité,* M. Lamertin, Bruxelles (1927).
10. Prigogine, I., et Defay, R., *Thermodynamique Chimique,* Mason Desoer Editions, Liege (1950).
11. Münster, A., *Chemische Thermodynamik,* Akademie-Verlag, Berlin (1969).
12. Mandel'shtam, L. I. and Leontovich, M. A., On the theory of sound absorption in liquids, *Zh. Eksp. Teor. Fiz.,* 7, 438 (1937).
13. Solov'ev, V. A., Thermodynamic theory of relaxation, in *Relaksationnye Yavleniya v Polimerach (Relaxation Phenomena in Polymers),* Bartenev, G. M. and Zelenev, Y. V., Eds., Khimiya, Moscow (1972), 129.
14. Davies, R. O. and Jones, G. O., Thermodynamic and kinetic properties of glasses, *Adv. Phys.,* 2, 370 (1953).
15. Staverman, A. J., Thermodynamic aspects of the glass-rubber transition, *Rheol. Acta,* 5, 283 (1966).
16. Breuer, H. and Rehage, G., Zur Thermodynamik der glassgen Erstarrung, *Kolloid Z. Z. Polym.,* 216/217, 159 (1967).
17. Gupta, P. K. and Moynihan, C. T., Prigogine-Defay ratio for systems with more than one order parameter, *J. Chem. Phys.,* 65, 4136 (1976).
18. Kovac, J., Equilibrium thermodynamics of the glass transition, *J. Phys. Chem.,* 85, 2060 (1981).
19. Nemilov, S. V., Stability criteria in the thermodynamic glass transition theory, *Fiz. i Khim. Stekla (Sov. J. Glass Phys. Chem.),* 13, 645 (1987).
20. Nemilov, S. V., Thermodynamic content of the Prigogine-Defay ratio and structural difference in the state of glass and liquid, in *Stekloobraznoje Sostoyanije,* Trudy 8 Vsesoyuznogo Soveshchania, October 28-31, 1986 *(Proc. 8th All-Union Conf. on the Vitreous State),* Porai-Koshitz, E. A., Ed., Nauka, Leningrad (1988), 15.
21. Storonkin, A. V., *Thermodynamics of Heterogeneous Systems,* parts 1 and 2, Izdat. LGU, Leningrad (1967), 42.
22. Müser, H. E. and Petersson, J., Thermodynamic theory of relaxation phenomena, *Fortschr. Phys.,* 19, 559 (1971).

23. Fichtenholz, G. M., *Course of Differential and Integral Calculum,* v. 1, Fizmatgiz, Moscow-Leningrad (1958), 424.
24. Fichtenholz, G. M., *Principles of Mathematical Analysis,* v. 2, Gostechizdat, Moscow (1957), 202.
25. Van't Hoff, J. H., Einfluss der Änderung der spezifischen Wärme auf die Umwandlungsarbeit, in *Boltzmann—Festschrift,* Verlag von Johann Ambrosius Barth, Leipzig (1904), 223.
26. Ehrenfest, P., Phasenumwandlungen im üblichen und erweiterten Sinn, klassifiziert nach den entsprechenden Singularitäten des thermodynamischen Potentials, *Proc. Acad. Amsterdam,* 36, 153 (1933).
27. Rao, C. N. and Rao, K. J., *Phase Transition in Solids,* McGraw-Hill, New York (1978), 20.
28. Smirnov, V. I., *Course of Higher Mathematics,* Gos. Izdat. Techn.-Teoret. Literat., Moscow-Leningrad, v. 3, part 1 (1949), 117.
29. Meixner, J., Discussion, in *Changement de Phases,* Compt. Rend. de la deuxième Réunion Annuelle tenue en commun avec la Commission de thermodyn. de l'Union Intern. de Physique, Paris, 2-7 juin 1952, Publ. Société de Chimie Physique, Paris (1952), 432.
30. DiMarzio, E. A., Validity of the Ehrenfest relation for a system with more than one order parameter, *J. Appl. Phys.,* 45, 4143 (1974).
31. Moynihan, C. T. and Lesikar, A. V., Comparison and analysis of relaxation processes at the glass transition temperature, *Ann. N.Y. Acad. Sci.,* 371, 151 (1981).
32. Gupta, P. K. and Haus, J. W., The Prigogine-Defay ratio for inhomogeneous systems with a single internal parameter, *J. Non-Cryst. Solids,* 24, 291 (1977).
33. Shultz, M. M. and Filippov, V. K., Thermodynamics of phase and chemical transformations in closed systems, *Zh. Vses. Khim. Ova. (Journal of D. I. Mendeleev All-Union Chemical Society),* 27, 485 (1982).
34. Nemilov, S. V., Bogdanov, V. N., Nikonov, A. M., Smerdin, S. N., Nedbai, A. I., and Borisov, B. F., Values of Prigogine-Defay Ratio for inorganic glassforming substances, *Fiz. i Khim. Stekla (Sov. J. Glass Phys. Chem.),* 13, 801 (1987).
35. Krol, D. M., Lyons, K. B., Brawer, S. A., and Kurkjian, C. R., High-temperature light scattering and the glass transition in vitreous silica, *Phys. Rev. B,* 33, 4196 (1986).
36. Hirai, N. and Eyring, H., Bulk viscosity of polymeric systems, *J. Polym. Sci.,* 37, 51 (1959).
37. Kauzmann, W., The nature of the glassy state and the behaviour of liquids at low temperatures, *Chem. Rev.,* 43, 218 (1948).
38. Douglas, R.W., The flow of glass, *J. Soc., Glass Technol.,* 33, 138 (1949).
39. Rubinstein, M., Resing, H. A., Reinecke, T. L., and Ngai, K. L., Nuclear spin-lattice relaxation in amorphous materials: B_2O_3, *Phys. Rev. Lett.,* 34, 1444 (1975).
40. Gibbs, J. H. and DiMarzio, E. A., Nature of the glass transition and the glassy state, *J. Chem. Phys.,* 28, 373 (1958).
41. DiMarzio, E. A. and Gibbs, J. H., Chain stiffness and the lattice theory of polymer phases, *J. Chem. Phys.,* 28, 807 (1958).
42. Gibbs, J. H., Nature of the glass transition and the vitreous state, in *Modern Aspects of the Vitreous State,* Mackenzie, J. D., Ed., Butterworths, London (1960), 152.

43. Vol'kenshtein, M. V., *Configurational Statistics of Polymer Chains,* Ed. by Acad. of Sci. of USSR, Moscow-Leningrad (1959).
44. Davies, R. O. and Lamb, J., Ultrasonic analysis of molecular relaxation process in liquids, *Q. Rev. London Chem. Soc.,* 11, 134 (1957).
45. Kneser, H. O., Allgemeine Theorie der Schallabsorption in Gasen und Flüssigkeiten, *Nuovo Cimento, Suppl.,* 7, ser. 9, 231 (1950).
46. Angell, C. A. and Rao, K. J., Thermodynamic and relaxational aspects of the glass transition from "Bond Lattice" model, in *Amorphous Materials,* Douglas, R. W. and Bryan Ellis, Eds., Wiley & Sons, London-New York (1972), 171.
47. Mazurin, O. V., Leko, V. K., Theory of chemical equilibria as a basis of interpretation of temperature-time dependencies of glassforming melts properties, *Fiz. i Khim. Stekla (Sov. J. Glass Phys. Chem.),* 9, 157 (1983).
48. Leidecker, H. W., Simmons, J. H., Litovitz, T. A., and Macedo, P. B., Degenerate exited state in the structure of B_2O_3, *J. Chem. Phys.,* 55, 2028 (1971).
49. Roe, R.-J., Thermodynamics of the glassy state with multiple order parameters, *J. Appl. Phys.,* 48, 4085 (1977).
50. Mozzi, R. L. and Warren, B. E., The structure of vitreous boron oxide, *J. Appl. Crystallogr.,* 3, 251 (1970).
51. Prabakar, S., Rao, K. J., and Rao C. N. R., ^{11}B NMR spectra and structure of boric oxide and alkali borate glasses, *Proc. R. Soc. London,* A429, 1 (1990).
52. Miyaka, M., Suzuki, T., Morikawa, H., Takagi, Y., and Marumo, F., Structural analysis of molten B_2O_3, *J. Chem. Soc., Faraday Trans.,* 80, part I , 1925 (1984).
53. Myuller (Müller, R. L.), The structure of solid glasses according to the data of electrical conductivity, *Izvest. Akad. Nauk SSSR, Ser. Fiz. (Bull. Acad. Sci. USSR,* 4, 607 (1940).
54. Smekal, A., Glaszustand und chemische Bindung Art, *Angew. Chem.,* 55, 235 (1942).
55. Syrkin, J. K. and Dyatkina, M. E., *Chemical Bond and Structure of Molecules,* Goskhimizdat, Moscow (1946), 406.
56. Meade, Ch., Hemley, R. J., and Mao, H. K., High-pressure X-ray diffraction of SiO_2 glass, *Phys. Rev. Lett.,* 69, 1387 (1992).
57. Yasui, I., Hasegawa, H., and Imaoka, M., X-ray diffraction study of the structure of silicate glasses. Part 1. Alkali metasilicate glasses, *Phys. Chem. Glasses,* 24, 65 (1983).
58. Imaoka, M., Hasegawa, H., and Yasui, I., X-ray diffraction study of the structure of silicate glasses. Part 2. Alkali disilicate glasses, *Phys. Chem. Glasses,* 24, 72 (1983).
59. Griscom, D. L., Borate glass structure, in *Borate Glasses. Structure, Properties, Applications,* Pye, L. D., Fréchette, V. D. and Kreidl N. J., Eds., Materials Science Research, v. 12, Plenum Press, New York-London (1978), 11.
60. Kozhina, E. L., Thermodynamic properties of the melts of potassium-silicate system, *Fiz. i Khim. Stekla (Sov. J. Glass Phys. Chem.),* 16, 679 (1990).
61. Emerson, J. F., Stallworth, P. E., and Bray, P. J., High-field ^{29}Si NMR studies of alkali silicate glasses, *J. Non-Cryst. Solids,* 113, 253 (1989).
62. Mackawa, H., Mackawa, T., Kawamura, K., and Yokokawa, T., The structural groups of alkali silicate glasses determined from ^{29}Si MAS NMR, *J. Non-Cryst. Solids.,* 127, 53 (1991).
63. Nikonov, A. M., Bogdanov, V. N., Nemilov, S. V., Shono, A. A., and Mikhailov, I. G., Structural relaxation in the melts of binary alkali-silicate glasses, *Fiz. i Khim. Stekla (Sov. J. Glass Phys. Chem.),* 8, 694 (1982).

64. Lomovskoi, V. A. and Bartenev, G. M., High-temperature relaxation spectra of borate glasses, *Sov. J. Glass Phys. Chem.* (Engl. transl. by Interperiodica Publ.), 19, 312 (1993).
65. Mozzi, R. L. and Warren, B. E., The structure of vitreous silica, *J. Appl. Crystallog.*, 2, 164 (1969).
66. Vukcevich, M. R., A new interpretation of the anomalous properties of vitreous silica, *J. Non-Cryst. Solids,* 11, 25 (1972).
67. Bartenev, G. M., *Superstrong and High-strong Inorganic Glasses,* Stroyizdat, Moscow (1974), 40.
68. Filipovich, V. N., On the nature of temperature dependence of structural rearrangements and fluctuations in melts and glasses, *Fiz. i Chim. Stekla (Sov. J. Glass Phys. Chem.),* 15, 322 (1989).
69. Rees, A. L. G., *Chemistry of the Defect Solid State,* Methuen & Co. Ltd., London (1954).
70. Filipovich, V. N., Problems of the theory of ionic conductivity of alkali-silicate glasses, *Wiss. Z. Friedrich-Schiller Univ. Jena, Math.-Naturwiss. Reihe,* 32, 317 (1983).
71. Richet, P. and Bottinga, Y., Glass transitions and thermodynamic properties of amorphous SiO_2, $NaAlSi_nO_{2n+2}$ and $KAlSi_2O_8$, *Geochim. Cosmochim. Acta,* 48, 453 (1984).
72. Visscher, P. B. and Logan, W. T., Stress as an order parameter for the glass transition, *Phys. Rev. B,* 42, 4779 (1990).
73. Leontovich, M., Relaxation in liquids and scattering of light, *J. Phys. (U.S.S.R.),* 4, 499 (1941).
74. Quentric, B., Local order in a liquid and the fine structure of vertical-horizontal light scattering, *Phys. Rev. A,* 15, 1304 (1977).
75. Angell, C. A. and Rao, K. J., Configurational excitations in condensed matter, and the "bond lattice" model for the liquid-glass transition, *J. Chem. Phys.,* 57, 470 (1972).
76. Goldstein, M., Viscous liquids and the glass transition. V. Sources of the excess specific heat of the liquid, *J. Chem. Phys.,* 64, 4767 (1976).
77. Moynihan, C. T., Macedo, P. B., Montrose, C. J., Gupta, P. K., DeBolt, M. A., Dill, J. F., Dom, B. E., Drake, P. W., Easteal, A. J., Elterman, P. B., Moeller, R. P., Sasabe, H., and Wilder, J. A., Structural relaxation in vitreous materials, *Ann. N. Y. Acad. Sci.,* 279, 15 (1976).
78. Zarzycki, J., Étude du reseau vitreux par diffraction des rayons X aux températures elevées, in *Trav. IVe Congrès Intern. du Verre,* Paris (1956), 323.
79. Zarzycki, J., Sur l'angle de la liasion Si–O–Si de la silica vitreuse et calculi de liasion Ge–O–Ge de l'oxyde de germanium vitreux ou liquide, *Verres et Refract.,* 11, 3 (1957).
80. Sokol'sky, V. E., Shovsky, V. A., Kazimirov, V. P., and Batalin, G. I., On the structure of vitreous silica at high temperatures, *Fiz. i Khim. Stekla (Sov. J. Glass Phys. Chem.),* 6, 517 (1980).
81. Nukui, A., Tagai, H., Mirikawa, H., and Iwai, S.-I., Structural study of molten silica by X-ray radial distribution analysis, *J. Am. Ceram. Soc.,* 61, 174 (1978).
82. Smart, R. M. and Glasser, F. P., Silicate anion constitution of lead silicate glasses and crystals, *Phys. Chem. Glasses,* 19, 95 (1978).
83. Kamiya, K., Yoko, T., Itoh, Y., and Sakka, S., X-ray diffraction study of Na_2O-GeO_2 melts, *J. Non-Cryst. Solids,* 79, 285 (1986).

84. Grimmer, A.-R. and Herms, G., B^{11} — NMR spectroscopische Untersuchung zur Temperaturabhängigkeit der Borkoordination in Natriumboratgläsern, *Z. Chem.,* 26, 452 (1986).
85. Derno, M., Herms, G., and Steil, H., X-ray study on the boron coordination in sodium borate glasses at high temperatures, *Phys. Status Solidi A,* 93, K5 (1986).
86. Geissberger, A. E. and Galeener, F. L., Raman studies of vitreous SiO_2 versus fictive temperature, *Phys. Rev. B,* 28, 3266 (1983).
87. Seifert, F. A., Mysen, B. O., and Virgo, D., Structure similarity of glasses and melts relevant to petrological processes, *Geochim. Cosmochim. Acta,* 45, 1879 (1991).
88. Furukawa, T. and White, W. B., Raman spectroscopic study of B_2O_3 and alkali borate glasses at high temperatures, *Phys. Chem. Glasses,* 21, 85 (1980).
89. Walrafen, G. E., Samanata, S. R., and Krishnan, P. N., Raman investigation of vitreous and molten boric oxide, *J. Chem. Phys.,* 72, 113 (1980).
90. Walrafen, G. E., Hokmabadi, M. S., Krishnan, P. N., Guha, S., and Munro, R. G., Low-frequency Raman scattering from vitreous and molten B_2O_3, *J. Chem. Phys.* 79, 3603 (1983).
91. Guha, S. and Walrafen, G. E., Low frequency Raman spectra of v-B_2O_3 as a function of temperature and pressure, *J. Chem. Phys.,* 80, 3807 (1984).
92. Johnson, P. A. V., Wright, A. C., and Sinclair, R. N., A neutron diffraction investigation of the structure of vitreous boron oxide, *J. Non-Cryst. Solids.,* 50, 281 (1982).
93. Malinovsky, V. K., Novikov, V. N., and Sokolov, A. P., Low-frequency Raman scattering in vitreous materials, *Fiz. i Khim. Stekla (Sov. J. Glass Phys. Chem.),* 15, 331 (1989).
94. Shuker, R. and Gammon, R. W., Low-frequency vibrational light scattering in viscous liquids, *J. Chem. Phys.,* 55, 4784 (1971).
95. Ritzhaupt, G. and Delvig, P., Low-temperature spectra of external modes in ionic glasses, *Chem. Phys. Lett.,* 21, 338 (1973).
96. Lorösch, J., Couzi, M., Pelous, J., Vicher, R., and Levasseur, A., Brillouin and Raman scattering study in borate glasses, *J. Non-Cryst. Solids,* 69, 1 (1984).
97. Kawamura, H., Fukumasu, F., and Hamada, Y., Low-frequency inelastic light scattering from As-S glasses, *Solid State Commun.,* 43, 229 (1982).
98. Malinovsky, V. K., Novikov, V. N., and Sokolov, A. P., Evaluation of the Rayleigh line wing and structural correlations for liquid-glass transition, *Phys. Lett. A,* 123, 19 (1987).
99. Fabelinsky, I. L., *Molecular Scattering of Light,* Nauka, Moscow (1965), 344.
100. Hassan, A. K., Torrell, L. M., and Borjesson, L., A. Raman study of B_2O_3 through the liquid-glass transition range, *J. Phys.* (Paris), 2, Colloque No. 2, Suppl. JP No. 10, C2-265 (1992).
101. Chempeney, D. C. and Sedgwick, D. F., Mössbauer scattering from supercooled liquids and glasses, *J. Phys. C,* 5, 1903 (1972).
102. Vasquez, A. and Flinn, P. A., Mössbauer effect investigation of the dynamics of ferrocene in liquid and glassy *o*-terphenyl, *J. Chem. Phys.,* 72, 1958 (1980).
103. Ellis, B. and McDonald, M. P., Broad-line NMR studies of α-D-glucose pentaacetate in crystalline and glassy states, *J. Non-Cryst. Solids,* 1, 186 (1968/1969).
104. Kramer, F., Müller-Warmuth, W., und Dutz, H., Magnetische Kernrezonanceuntersuchungen an Gläsern in Transformationsbereich, *Glastech. Ber.,* 46, 191 (1973).

105. Startsev, Y. K., Kluev, V. P., and Vostrikova, M. C., Determination of glass transition temperatures by simultaneously regulated dependencies of thermal expansion and conductivity, *Fiz. i Khim. Stekla (Sov. J. Glass Phys. Chem.),* 4, 278 (1978).
106. Stebbins, J. F., Effects of temperature and composition on silicate glass structure and dynamics: Si-29 NMR results, *J. Non-Cryst. Solids,* 106, 359 (1988).
107. Phillips, W. A., Buchenau, U., Nücker, N., Dianoux, A.-J., and Petry, W., Dynamics of glassy and liquid selenium, *Phys. Rev. Lett.,* 63, 2381 (1989).
108. Buchenau, U. and Zorn, R., A relation between fast and slow motions in glassy and liquid selenium, *Europhys. Lett.,* 18, 523 (1992).
109. Saunders, G. A., Phonon anharmonicity near the melting point and the glass transition, *Philos. Mag., B,* 59, 179 (1989).
110. Nemilov, S. V. and Mikhailov, V. V., Thermospectrophotometrical study of structural changes in vitreous $Co(PO_3)_2$ at the transition from solid to liquid, *Fiz. i Khim. Stekla (Sov. J. Glass Phys. Chem.),* 6, 241 (1980).
111. Mikhailov, V. V. and Nemilov, S. V., Investigation of Co^{2+} spectra in haloid-containing vitreous matrices at the transition from solid to liquid, *Fiz. i Khim. Stekla (Sov. J. Glass Phys. Chem.),* 7, 444 (1981).
112. Nemilov, S. V. and Mikhailov, V. V., Thermospectrophotometrical study of structural changes in borate and silicate glasses at the transition from solid to liquid, *Fiz. i Khim. Stekla (Sov. J. Glass Phys. and Chem.),* 7, 669 (1981).
113. Jäckle, J., Residual entropy in glasses and spin glasses, *Physica BC,* 127, 79 (1984).
114. Simon, F., Fünfundzwanzig Jahre Nernstschen Wärmesatz, *Ergeb. Exakten Naturwiss.,* 9, 222 (1930).
115. Nemilov, S. V., Thermodynamics of glassy state and the Nernst Heat Theorem, *Fiz. i Khim. Stekla (Sov. J. Glass Phys. Chem.),* 15, 305 (1989).
116. Wilks, J., *The Third Law of Thermodynamics,* Oxford Univ. Press, Oxford, UK (1961), 45.
117. Blumberg, B. Ya. *Introduction into Physical Chemistry of Glass,* Gos. nauchno-techn. isdat. chim. prom., Leningrad (1939), 33.
118. Obald, A. and Newton, R., The heat capacity of supercooled liquid glycerol, *J. Am. Chem. Soc.,* 59, 2495 (1937).
119. Gonchukova, N. O., Investigation of the entropy of vitreous and crystalline lithium acetate in connection with the Kauzmann Paradox, *Fiz. i Khim. Stekla (Sov. J. Glass Phys. Chem.),* 8, 429 (1982).
120. Zelinski, B. J. J., Yinnon, H., and Uhlmann, D. R., The Kauzmann paradox revisited, *Glastech. Ber.,* 56K, 822 (1983).
121. Angell, C. A., MacFarlane, D. K., and Oguni, M., The Kauzmann paradox, metastable liquids, and ideal glasses, *Ann. N. Y. Acad. Sci.,* 484, 241 (1986).
122. DiMarzio, E. A., Equilibrium theory of glasses, *Ann. N. Y. Acad. Sci.,* 371, 1 (1981).
123. Kerner, R., Geometrical approach to the glass transition problem, *J. Non-Cryst. Solids* 71, 19 (1985).
124. Stillinger, F. H., Supercooled liquids, glass transition, and the Kauzmann paradox, *J. Chem. Phys.,* 88, 7818 (1988).
125. Milchev, A. and Gutzow, I., Temperature dependence of the configurational entropy of undercooled melts and the nature of glass transition, *J. Macromol. Sci., Phys.,* B21, 583 (1982).

126. Murthy, S. S. N., Kauzmann paradox and the structure of glass, *J. Mol. Liq.*, 44, 119 (1990).
127. Woodcock, L. V., Glass transition in the hard-sphere model and Kauzmann's paradox, *Ann. N. Y. Acad. Sci.*, 371, 274 (1981).
128. Vashishta, P., Kalia, R. K., Rino, J. P., and Ebbsjö, I., Interaction potential for SiO_2: A molecular-dynamics study of structural correlations, *Phys. Rev. B*, 41, 12197 (1990).
129. Nemilov, S. V., Relationship between reduced thermodynamic functions of vitreous systems at absolute Zero, *Fiz. i Khim. Stekla (Sov. J. Glass Phys. Chem.)*, 7, 575 (1981).
130. Chang, S. S., Horman, J. A., and Bestul, A. B., Heat capacities and relative thermal data for diethyl phthalate crystal, glass, and liquid to 360 K, *J. Res. Nat. Bur. Stand.*, 71A, 293 (1967).
131. Nemilov, S. V., Entropy and the valence structure of glass, *Fiz. i Khim. Stekla (Sov. J. Glass Phys. Chem.)*, 2, 97 (1976).
132. Nemilov, S. V., Thermodynamic functions of nonequilibrium disordered systems at absolute zero and the nature of the vitreous state, *Fiz. i Khim. Stekla (Sov. J. Glass Phys. Chem.)*, 8, 11 (1982).
133. Westrum, E., Grenier, D. M., The heat capacities and thermodynamic properties of crystalline and vitreous anhydrous sodium tetraborate, *J. Am. Chem. Soc.*, 79, 1799 (1957).
134. Chang, S. S. and Bestul, A. B., Heat capacity of selenium crystal (trigonal), glass, and liquid from 5 to 360 K, *J. Chem. Thermodyn.*, 6, 325 (1974).
135. Chang, S. S. and Bestul, A. B., Heat capacity and thermodynamic properties of *o*-terphenyl crystal, glass, and liquid, *J. Chem. Phys.*, 56, 503 (1972).
136. Parks, G. S., Thomas, S. B., and Light, D. W., XII. Some new heat capacity data for organic glasses. The entropy and free energy of dl-lactic acid, *J. Chem. Phys.*, 4, 64 (1936).
137. Wunderlich, B. and Baur, H., Heat capacities of linear high polymers, *Fortschr. Hochpolym. Forsch.*, 7, 151 (1970).
138. Einstein, A., Die Plancksche Theorie der Strahlung und die Theorie der spezifischen Wärme, *Ann. Phys.*, 22, 180 (1907).
139. Debye, P., Zur Theorie der spezifischen Wärmen, *Ann. Phys.*, (4 Folge), 39 789 (1912).
140. Tarassov, V. V., *New Problems of the Physics of Glass*, Gosstroyizdat, Moscow (1959).
141. Tarassov, V. V. and Yunitsky, G. A., Heat capacity theory of chain-layer structures, *Zh. Fiz. Khim. (Russ. J. Phys. Chem.)*, 39, 2077 (1965).
142. Deltour, J. and Kartheuser, E., Numerical computation of the thermodynamic properties of the ordered diatomic linear chain, *Bull. Soc. Roy. Sci. Liege*, 33, 801 (1964).
143. Deltour, J. and Kartheuser, E., Thermodynamic properties of the ordered diatomic linear chain, *Physica*, 31, 269 (1965).
144. Soga, N., Three-band theory and elastic moduli of glass, *J. Non-Cryst. Solids*, 52, 365 (1982).
145. Hirao, K. and Soga, N., Heat capacity and chemical bond strength of oxide glasses, *J. Ceram. Soc. Jpn*, 97, 359 (1989).
146. Pohl, R. O., Low-temperature specific heat of glasses, in *Amorphous Solids. Low-Temperature Properties*, Phillips, W. A., Ed., Springer-Verlag, Berlin-Heidelberg-New York (1981), 27.

147. Phillips, W. A., Introduction, in *Amorphous Solids. Low-Temperature Properties,* Phillips, W. A., Ed., Springer-Verlag, Berlin-Heidelberg-New York (1981), 1.
148. Buchenau, U., Soft localized vibrations in glasses and undercooled liquids, *Philos. Mag. B,* 65, 303 (1992).
149. Buchenau, U., Dynamics of glassy materials as studied by neutrons, *Physica B,* 174, 131 (1991).
150. Klinger, M. I., Soft atomic configurations, medium-range order, and large-scale structural transformations in glasses, *Sov. Glass Phys. Chem.* (Engl. transl. by Interperiodica Publ.), 18, 353 (1992).
151. Richet, P. and Bottinga, Y., Thermochemical properties of silicate glasses and liquids: A review, *Rev. Geophys.,* 24, 1 (1986).
152. Mazurin, O. V., Analysis of temperature dependencies of some properties of high-viscous liquids basing on the modern concept on structure relaxation, *Wiss. Z. Friedrich-Schiller Univ. Jena, Math.-Naturwiss. Reihe,* 28, 507 (1979).
153. Schnaus, U. E., Moynihan, C. T., Gammon, R. W., and Macedo, P. B., The relation of the glass transition temperature to vibrational characteristics of network glasses, *Phys. Chem. Glasses,* 11, 213 (1970).
154. Wunderlich, B., Study of the change in specific heat of monomeric and polymeric glasses during the glass transition, *J. Phys. Chem.,* 64, 1052 (1960).
155. Wunderlich, B. and Jones, L. D., Heat capacities of solid polymers, *J. Macromol. Sci., Phys.,* 3, 67 (1969).
156. Varma-Nair, M., Pan, R., and Wunderlich, B., Heat capacities and entropies of linear, aliphatic polyesters, *J. Polymer Sci., Part B,* 29, 1107 (1991).
157. Nemilov, S. V., Relation of the excess thermodynamic functions of vitreous As_2S_3 and As_2Se_3 at absolute zero to their structural features, *Fiz. i Khim. Stekla (Sov. J. Glass Phys. Chem.),* 8, 465 (1982).
158. Nemilov, S. V. and Muratov, A. V., Heat capacity and excess thermodynamic functions at absolute zero of vitreous $Na_2O{\cdot}2SiO_2$ and $Na_2O{\cdot}SiO_2$ as connected with the structure of soda-silica glasses, *Fiz. i Khim. Stekla (Sov. J. Glass Phys. Chem.),* 9, 589 (1983).
159. Openshaw, R. E., Hemingway, B. E., Robie, R. A., Waldbaum, D. R., and Krupka, K. M., The heat capacities at low temperatures and entropies at 298.15 K of low albite, analbite, microcline, and high sanidine, *U.S. Geol. Survey Journ. Res.,* 4, 195 (1976).
160. Nemilov, S. V., Structural randomness entropy of low-molecular-weight organic liquids, glasses and glassy crystals and its relation to the structure of the molecules, *Fiz. i Khim. Stekla (Sov. J. Glass Phys. Chem.),* 3, 423 (1977).
161. Lebedev, B. V., Rabinovich, I. B., Yevstropov, A. A., and Rykhl'evskaya, M. S., Thermodynamic properties of N-(β-trimethylsilylethyl)trimethyleneimine in the region of 7-305 K, in *Thermodynamics of Organic Compounds,* 4, No. 43, 27 (1975).
162. Milov, V. I., Lebedev, B. V., Kiparisova, Ye. G., Bykova, T. A., and Sladkov, A. M., Thermodynamics of diphenyldiethinylsilane in the region of 10-320 K, in *Thermodynamics of Organic Compounds,* Rabinovich, I. B., Ed., Gorky State University Publishers, Gorky, 6, 13 (1977).
163. Babinkov, A. G., Rabinovich, I. B., and Mikhaleva, L. A., Heat capacity of oligoethylenebutyleneglycole adipinate in the range of 6-330 K, in *Thermodynamics of Organic Compounds,* Rabinovich, I. B., Ed., Gorky State University Publishers, Gorky, 3, (1982).

164. Jafarov, O. I. and Kuliyev, A. M., Heat capacity and thermodynamic properties of 1,1,3,3 tetraethyl-5,5-diphenyl-cyclotrisiloxane in the range of 5-315 K, *Zh. Fiz. Khim. (Russ. J. Phys. Chem.),* 61, 606 (1987).
165. Nistratov, V. P., Rabinovich, I. B., Sheiman, M. S., Smirnova, N. N., Zeljaev, I. A., and Feschenko, I. A., Heat capacity and thermodynamic characteristics of triethylbismuth, *Zh. Fiz. Khim. (Russ. J. Phys. Chem.),* 63, 1779 (1989).
166. Lebedev, B. V., Yevstropov, A. A., and Kiparisova, Ye. G., Thermodynamics of undecanolactone, tridecanolactone, and pentadecanolactone from 0 to 340 K, *J. Chem. Thermodyn.,* 13, 1185 (1981).
167. Gujrati, P. D. and Goldstein, M., Viscous liquids and the glass transition. 9. Nonconfigurational contributions to the excess entropy of disordered phases, *J. Phys. Chem.,* 84, 759 (1980).
168. Lasjaunias, J. C., Penn, A., Ravex, A., and Vandrope, M., Effect of thermal treatments on the low temperature specific heat of B_2O_3 and SiO_2 glasses, *J. Phys. (Paris), Lett.,* 41, L-131 (1980).
169. Johari, G. P., Secondary relaxation and the properties of glasses and liquids, in *Molecular Dynamics and Relaxation Phenomena in Glasses,* Dorfmüller, Th. and Williams, G., Eds., Springer-Verlag, Berlin-Heidelberg-New York-London-Paris-Tokyo (1987), 90.
170. Privalko, V. P., Excess entropies and related quantities in glass-forming liquids, *J. Phys. Chem.,* 84, 3307 (1980).
171. Haida, O., Suga, H., and Seki, S., Calorimetric study of the glassy state. XI. Plural glass transition phenomena in cyclohexene, *Bull. Chem. Soc. Jpn,* 50, 802 (1977).
172. Haida, O., Suga, H., and Seki, S., Calorimetric study of the glassy state. XII. Plural glass-transition phenomena of ethanol, *J. Chem. Thermodyn.,* 9, 1138 (1977).
173. Simon, F. E., The Third Law of Thermodynamics. An historical survey, in *Year Book Phys. Soc. London,* Ed. by Phys. Soc. London (1956), 1.
174. Jäckle, J., On the glass transition and the residual entropy of glasses, *Philos. Mag. B,* 44, 533 (1981).
175. Stishov, S. M., Entropy, disorder and melting, *Usp. Fiz. Nauk (Progr. in Phys. Sci. USSR),* 154, 93 (1988).
176. Palmer, R. G., Broken ergodicity, *Adv. Phys.,* 31, 669, (1982).
177. Rostiashvili, V. G., Irzak, V. I., and Rosenberg, B. A., *Glass Transition in Polymers,* Khimiya, Leningrad (1987) 192.
178. Forster, D., *Hydrodynamic Fluctuations, Broken Symmetry, and Correlation Functions,* Reading: Benjamin, London, (1975).
179. Thirumalai, D., Mountain, R. D., and Kirkpatrick, T. R., Ergodic behavior in supercooled liquids and in glasses, *Phys. Rev. A,* 39, 3563 (1989).
180. Palmer, R. G., Parallels and contrasts between glass and spin glass, *Ann. N. Y. Acad. Sci.,* 484, 109 (1986).
181. Goldstein, M., Viscous liquids and glass transition: a potential barrier picture, *J. Chem. Phys.,* 51, 3728 (1969).
182. Stillinger, F. H., Relaxation behavior in atomic and molecular glasses, *Phys. Rev. B,* 41, 2409 (1990).
183. Lebedev, A. A., On polymorphism and annealing of glass, *Trudy Gos. Opt. Inst. (Proc. of S. I. Vavilov State Optical Institute),* 11, No. 10, 1 (1921).
184. Pauling, L. and Tolman, R. C., The entropy of supercooled liquids at the absolute Zero, *J. Am. Chem. Soc.,* 47, 2148 (1925).

185. Kelley, K. K., The heat capacities of ethyl and hexyl alcohols from 16 °K to 298 °K and the corresponding entropies and free energies, *J. Am. Chem. Soc.*, 51, 779 (1929).
186. Gutzow, I., Über die Nullpunktsentropie des Glaszustandes, *Z. Phys. Chem.*, 221, 153 (1962).
187. Bell, R. J. and Dean, P., The configurational entropy of vitreous silica in the random network theory, *Phys. Chem. Glasses*, 9, 125 (1968).
188. Narten, A. N., Diffraction pattern and structure of noncrystalline BeF_2 and SiO_2 at 25°C, *J. Chem. Phys.*, 56, 1905 (1972).
189. Da Silva, J. R. G., Pinatti, D. G., Anderson, C. E., and Rudee, M. L., A refinement of the structure of vitreous silica, *Philos. Mag.*, 31, 713 (1975).
190. Dupree, E. and Pettifer, R. F., Determination of the Si–O–Si bond angle distribution in vitreous silica by magic angle spinning NMR, *Nature (London)*, 308, No. 5959, 523 (1984).
191. Gladden, L. F., Carpenter, T. A., and Elliott, S. R., ^{29}Si MAS NMR studies of the spin-lattice relaxation time and bond-angle distribution in vitreous silica, *Philos. Mag. B*, 53, L 81 (1986).
192. Thathachari, Y. T. and Tiller, W. A., Steric origin of the silicon- oxygen angle distribution in silica, *J. Appl. Phys.*, 53, 8615 (1982).
193. Bell, R. J. and Dean, P., The structure of vitreous silica: Validity of the random network theory, *Philos. Mag.*, 25, 1381 (1972).
194. Gladden, L. F., Medium-range order in v-SiO_2, *J. Non-Cryst. Solids*, 119, 318 (1990).
195. Mosseri, R. and Sadoc, J. F., From polytopes to non-crystalline structures: the iterative flattering method, *J. Non-Cryst. Solids*, 75, 115 (1985).
196. Morris, R. C., Disclination-dislocation model of metallic glass structures, *J. Appl. Phys.*, 50, 3250 (1979).
197. Zallen, R., *The Physics of Amorphous Solids*, Wiley & Sons, New York (1983), 49.
198. Nelson, D. R., Order, frustration and defects in liquids and solids, *Phys. Rev. B*, 28, 5517 (1983).
199. Lichatchev, V. A. and Mikhailin, A. I., Displanations in glasses, *Fiz. i Khim. Stekla (Sov. J. Glass Phys. Chem.)*, 14, 161 (1988).
200. Lichatchev, V. A. and Shudegov, V. E., Disclinational model of vitreous silica structure, *Fiz. i Khim. Stekla (Sov. J. Glass Phys. Chem.)*, 15, 508 (1989).
201. River, N. and Duffy, D. M., On the topological entropy of atomic liquids and the latent heat of fusion, *J. Phys. C*, 15, 2867 (1982).
202. Gupta, P. K. and Cooper, A. R., Topological disordered networks of rigid polytopes: applications, in *Physics of Non-Crystalline Solids*, Pye, L. D., LaCourse, W. C., and Stevens, H. J., Eds., Taylor & Francis, London-Washington (1992) 72.
203. Kob, W. and Schilling, R., Cooling rate dependence of the residual energy of a one-dimensional configurational glass model, *Z. Phys. B*, 68, 245 (1987).
204. Gaskell, P. H. and Tarrant, I. D., Refinement of random network model for vitreous silicon dioxide, *Philos. Mag. B*, 42, 265 (1980).
205. Nemilov, S. V., On the conformity of the results of structural and thermodynamic studies of vitreous silica, *Fiz. i Khim. Stekla (Sov. J. Glass Phys. Chem.)*, 8, 385 (1982).
206. Bock, J. and Gouq-Jen Su, Interpretation of the infrared spectra of fused silica, *J. Am. Ceram. Soc.*, 53, 69 (1970).

207. Howitt, D. G. and McElfresh, D. K., A calculation of the free energies of some structural models for vitreous silica, *Phys. Chem. Glasses,* 23, 107 (1982).
208. Lindenmeyer, P. H., The inhomogenous structure of glass, *Polym. Eng. Sci.,* 21, 958 (1981).
209. Fujita, F. E., Statistical thermodynamical study of medium range order, *J. Non-Cryst. Solids,* 106, 286 (1988).
210. Vainas, J. and Almond, D. P., A geometrical rationalization of frozen-in enthalpy and frozen-in entropy in glasses, *J. Phys.: Cond. Matter,* 4, L 85 (1992).
211. Evstropyev, K. S. and Porai-Koshitz, E. A., Discussion on the modern state of the crystallite hypothesis of glass structure, *J. Non-Cryst. Solids,* 11, 170 (1972).
212. Yeh, G. S., Current concepts of morphology of amorphous polymers, *Vysokomol. Şoedin. [Polym. Sci. (USSR)],* 21A, 2433 (1979).
213. Červinka, L., Medium-range ordering in amorphous solids. Example: polycarbonates, *J. Non-Cryst. Solids,* 75, 638 (1985).
214. Novikov, V. N. and Sokolov, A. P., A correlation between low-energy vibrational spectra and the first sharp diffraction peak in chalcogenide glasses, *Solid State Commun.,* 77, 243 (1991).
215. Wright, A. C., Neutron and X-ray amorphography, *J. Non-Cryst. Solids* 106, 1 (1988).
216. Gerber, T. and Himmel, B., Röntgenbeugungsuntersuchungen zur Struktur von glassigen und amorphem SiO_2, *Wiss. Z. Wilhelm-Pieck-Univ. Rostock, Naturwiss. R.,* 36, 14 (1987).
217. Brauer, G. and Boden, G., Ordering phenomena in silica glasses studied by positron annihilation, *Diffusion and Defect Data,* 53-54, 173 (1987).
218. Nukui, A., Tagai, H., Morikawa, H., and Iwai, S.-J., Structural study of molten silica by an X-ray radial distribution analysis, *J. Am. Ceram. Soc.,* 61, 174 (1978).
219. Golubkov, V. V., The structure and structural transformations in vitreous SiO_2, *Sov. J. Glass Phys. Chem.* (Engl. transl. by Interperiodica Publ.), 18, 28 (1992).
220. Arndt, J. and Stoffler, D., Anomalous changes in some properties of silica glass densified at very high pressures, *Phys. Chem. Glasses,* 10, 117 (1969).
221. Grimsditch, M., Polymorphizm in amorphous SiO_2, *Phys. Rev. Lett.,* 52, 2379 (1984).
222. Velde, B., and Couty, R., High-pressure infrared spectra of silica glass and quartz, *J. Non-Cryst. Solids,* 94, 238 (1987).
223. Porai-Koshitz, E. A., On short- and medium-range order in glass structure, *Sov. J. Glass Phys. Chem.* (Engl. transl. by Interperiodica Publ.) 18, 361 (1992).
224. Denbigh, K. G., *The Thermodynamics of the Steady State,* Methuen, London, Wiley & Sons, New York (1951), 46.
225. Nemilov, S. V., Entropy production in irreversible processes of glass structure stabilization and a thermodynamic validation of the memory effect, *Fiz. i Khim. Stekla (Sov. J. Glass Phys. Chem.),* 11, 146 (1985).
226. Haase, R., Survey of fundamental laws, in *Physical Chemistry. An Advanced Treatise,* v. I, *Thermodynamics,* Jost, E. W., Ed., Academic Press, New York-London (1971), 2.
227. DeBolt, M. A., Easteal, A. J., Macedo, P. B., and Moynihan, C. T., Analysis of structural relaxation in glass using rate heating data, *J. Am. Ceram. Soc.,* 59, 16 (1976).

228. Vol'kenshtein, M. V., Structural and kinetic characteristics of glassy state, in *The Structure of Glass* (Proc. 3 All-Union Conf. on the Glassy State, Leningrad, November 16-20, 1959), Porai-Koshitz, E. A., Ed., (engl. transl. llvarov, B., Ed.), Consultants Bureau, New York (1960), 50.
229. Mazurin, O. V., *Glass Transition,* Nauka, Leningrad (1986), 23.
230. Vuks, M. F., *Light Scattering in Gases, Liquids and Solutions,* Izdat. LGU, Leningrad (1977), 134.
231. Vasilevskaya, T. A., Golubkov, V. V., Titov, A. P., and Porai-Koshitz, E. A., Investigation of structure of vitreous B_2O_3 by X-ray scattering under small angles, in *Stekloobraznoye Sostoyaniye,* Trudy 8 Vsesoyuznogo Soveschania, October 28-31, 1986 (Proc. 8th All-Union Conf. on the Vitreous State), Porai-Koshitz, E. A., Ed., Nauka, Leningrad (1988), 43.
232. Bokov, N. A., Temperature dependence of visible light scattering at the transition of glass into metastable liquid, *Fiz. i Khim. Stekla (Sov. J. Glass Phys. Chem.),* 17, 839 (1991).
233. Florinskaya, V. A., Variations of refractive index and dispersion in optic glass under heat treatment, *Trudy Gos. Opt. Inst. (Proc. of S. I. Vavilov State Optical Institute),* 19, No. 131, 3 (1950).
234. Ritland, H. N., Limitations of the fictive temperature concept, *J. Am. Ceram. Soc.,* 39, 403 (1956).
235. Spinner, S. and Napolitano, A., Further studies in the annealing of a borosilicate glass, *J. Res. Nat. Bur. Stand.,* 70A, 147 (1966).
236. Macedo, P. B. and Napolitano, A., Effects of a distribution of volume relaxation times in the annealing of BSC glass, *J. Res. Nat. Bur. Stand.,* 71A, 231 (1967).
237. Boesch, L., Napolitano, A., and Macedo, P. B., Spectrum of volume relaxation times in B_2O_3, *J. Am. Ceram. Soc.,* 53, 148 (1970).
238. Mizouchi, N. and Cooper, A. R., Relaxation of volume and index of refraction in pressure-compacted B_2O_3 glass, *J. Am. Ceram. Soc.,* 56, 320 (1973).
239. Nemilov, S. V., Structural-kinetic interpretation of the memory effect in glasses upon a stepwise change in stabilization temperature, *Fiz. i Khim. Stekla (Sov. J. Glass Phys. Chem.),* 12, 293 (1986).
240. Napolitano, A. and Macedo, P. B., Spectrum of relaxation times in GeO_2 glass, *J. Res. Nat. Bur. Stand.,* 72A, 425 (1968).
241. Macedo, P. B. and Litovitz, T. A., Ultrasonic viscous relaxation in molten boron trioxide, *Phys. Chem. Glasses,* 6, 69 (1965).
242. Kovacs, A. J., Transition vitreuse dans polymères amorphs: Etude phénomenologique, *Fortschr. Hochpolym. Forsch.,* 3, 394 (1963).
243. Cooper, A. R., Trigonal networks and memory effects in vitreous B_2O_3, *Fiz. i Khim. Stekla (Sov. J. Glass Phys. Chem.),* 2, 24 (1976).
244. De Bast, J. and Gilard, P., Reologie du verre sans contraints dans l'intervalle transformations, *Rech. Trav. Cent. Tech. Sci. Ind. Belge Verre,* 36, 192 (1969).
245. Palmer, R. O., Slein, D. L., Abrahams, E., and Anderson, P. W., Models to hierarchically constrained dynamics for glassy relaxation, *Phys. Rev. Lett.,* 53, 958 (1984); Errata: 54, 1965E (1985).
246. Kohlrausch, R., Theorie des elektrischen Rückstands in der Leidener Flasche, *Pogg. Ann. Physik und Chemie,* 91, 179 (1854).
247. Williams, G. and Watts, D. C., Non-symmetric dielectric relaxation behaviour arising from a simple empirical decay function, *Trans. Faraday Soc.,* 66, 80 (1970).

248. Gardon, R. and Narayanaswamy, O. S., Stress and volume relaxation in the annealing of flat glass, *J. Am. Ceram. Soc.,* 53, 380 (1970).
249. Kovacs, A. J., A multiparameter approach for structural recovery of glasses and its implication for their physical properties, *Ann. N. Y. Acad. Sci.,* 371, 38 (1981).
250. Gupta, P. K., Fictive pressure effects in structural relaxation, *J. Non-Cryst. Solids,* 102, 231 (1988).
251. Moynihan, C. T. and Gupta, P. K., The order parameter model for structural relaxation in glass, *J. Non-Cryst. Solids,* 29, 143 (1978).
252. Moynihan, C. T., Theory and experimental investigation of structural relaxation in glass, *Wiss. Z. Friedrich-Schiller-Univ. Jena, Math.-Naturwiss. Reihe,* 28, 493 (1979).
253. Lesikar, A. V. and Moynihan, C. T., Some relations connecting volume and enthalpy relaxation in the order parameter model of liquids and glasses, *J. Chem. Phys.,* 72, 6422 (1980).
254. Berg, J. I. and Cooper, A. R., Linear nonequilibrium thermodynamic theory of glass transition kinetics, *J. Chem. Phys.,* 68, 4481 (1978).
255. Fujimori, H., Adachi, Y., and Oguni, M., Temperature-jump method for characterization of structure fluctuations and irreversible relaxation processes in liquids and glasses, *Phys. Rev. B.,* 46, 14501 (1992).
256. Birge, N. O., Jeong, Y. H., and Nagel, S. R., Specific heat and ultrasonics as dynamic probes of the glass transition, *Ann. N.Y. Acad. Sci.,* 484, 101 (1986).
257. Bartenev, G. M., *Structure and Mechanical Properties of Inorganic Glasses,* Stroyizdat, Moscow (1966), chap. 1.
258. Born, M. and Green, H. S., *A General Kinetic Theory of Liquids,* Cambridge Univ. Press, Cambridge, UK (1949), chap. II.
259. Barrer, R. M., Permeability in relation to viscosity and structure of rubber, *Trans. Faraday Soc.,* 38, 322 (1942).
260. Bernal, J. D., An attempt at a molecular theory of liquid structure, *Trans. Faraday Soc.,* 33, 27 (1937).
261. Nemilov, S. V., Genesis of the vitreous state: Three approaches to the problem, *Sov. J. Glass Phys. Chem.* (Engl. transl. by Interperiodica Publ.), 18, 341 (1992).
262. Hopkins, I. L. and Kurkjian, C. R., Relaxation spectra and relaxation processes in solid polymers and glasses, in *Physical Acoustics,* v. II, part B, *Properties of Polymers and Nonlinear Acoustics,* Mason, W. P., Ed., Academic Press, New York-London (1965), 91.
263. Mazurin, O. V. and Potseluyeva, L. N., Determination of glass transition temperatures by temperature dependencies of viscosity of glassforming melts, *Fiz. i Khim. Stekla (Sov. J. Glass Phys. Chem.),* 4, 570 (1978).
264. Mazurin, O. V., Startsev, Y. K., and Potseluyeva, L. N., Calculation of the time of high-viscous liquid reaching the state of metastable equilibrium, *Fiz. i Khim. Stekla (Sov. J. Glass Phys. Chem.),* 4, 675 1978.
265. Mazurin, O. V., Startsev, Y. K., and Potseluyeva, L. N., Investigation of temperature dependencies of viscosities for some glasses at constant structural temperature, *Fiz. i Khim. Stekla (Sov. J. Glass Phys. Chem.),* 5, 82 (1979).
266. Nemilov, S. V., Kantchieva, O. N., and Komarova, N. V., The viscosity, elastic properties and cohesion energy of glasses in the systems Na_2O–GeO_2–TiO_2 and K_2O–GeO_2–TiO_2, *Fiz. i Khim. Stekla (Sov. J. Glass Phys. Chem.),* 12, 172 (1986).
267. Barrer, R. M., The viscosity of pure liquids, *Trans. Faraday Soc.,* 39, 48 (1943).

268. Napolitano, A. and Hawkins, E. G., Viscosity of a standard soda-lime-silica glass, *J. Res. Nat. Bur. Stand.*, 68A, 439 (1964).
269. Macedo, P. B. and Litovitz, T. A., Inadequacies of viscosity theories for B_2O_3, *J. Chem. Phys.*, 49, 1887 (1968).
270. Uhlmann, D. R., Viscous flow in glassforming liquids, in *Amorphous Materials,* Douglas, R. W. and Ellis, B., Eds., Wiley-Interscience, London-New York-Sidney-Toronto (1972), 205.
271. Oldekop, W., Theoretische Betrachtungen über die Zähigkeit von Gläsern, *Glastech. Ber.*, 30, 8 (1957).
272. Nemilov, S. V., On the relationship between the activation entropy of viscous flow, the heat capacity and valence structure of glass, *Zh. Prikl. Khim. [J. Appl. Chem. (USSR)],* 37, 293 (1964).
273. Nemilov, S. V., Viscosity and structure of glass, in *The Structure of Glass,* (Proc. 4th All-Union Conference on the Glassy State, Leningrad, March 16-21, 1964), v. 6, part I, Porai-Koshitz, E. A., Ed., (Engl. transl. Uvarov, B., Ed.), Consultants Bureau, New York (1966), 58.
274. Nemilov, S. V., An analysis of energetic activation parameters and the nature of viscous flow of inorganic glasses, in *Uspekhi Reologii Polimerov* (Progress in the Rheology of Polymers), Vinogradov, G. V., Ed., Khimiya, Moscow (1970), 241.
275. Nemilov, S. V., Viscous flow of glasses in relation to their structure: Application of the theory of rate processes, *Sov. J. Glass Phys. Chem.* (Engl. transl. by Interperiodica Publ.), 18, 1 (1992).
276. Laughlin, W. T. and Uhlmann, D. R., Viscous flow in simple organic liquids, *J. Phys. Chem.*, 76, 2317 (1972).
277. Angell, C. A., Spectroscopy simulation and scattering, and the medium range order problem in glass, *J. Non-Cryst. Solids,* 73, 1 (1985).
278. Wong, J. and Angell, C. A., *Glass Structure by Spectroscopy,* Marcel Dekker, New York (1976), chap. 1.
279. Angell, C. A. and Sichina, W., Thermodynamics of the glass transition: empirical aspects, *Ann. N.Y. Acad. Sci.*, 276, 53 (1976).
280. Angell, C. A., Structure instability and relaxation in liquid and glassy phases near their fragile liquid limit, *J. Non-Cryst. Solids,* 102, 205 (1988).
281. Angell, C. A., Perspective on the glass transition, *J. Phys. Chem. Solids.*, 49, 863 (1988).
282. Angell, C. A., Relaxation in liquids, polymers and plastic crystals — strong/fragile pattern and problems, *J. Non-Cryst. Solids,* 131-133, 13 (1991).
283. Sanditov, D. S. and Bartenev, G. M., *Physical Properties of Disordered Structures,* Nauka, Novosibirsk (1982), chap. 2.
284. Batschinski, A. I., Über die innere Reibung der Flüssigkeiten, *Z. Phys. Chem.*, 84, 643 (1913).
285. Doolittle, A. K., Studies in newtonian flow. II. The dependence of the viscosity of liquids on free space, *J. Appl. Phys.*, 22, 1471 (1951).
286. Williams, M. L., Landel, R. F., and Ferry, J. D., The temperature dependence of relaxation mechanisms in amorphous polymers and other glass-forming liquids, *J. Am. Chem. Soc.*, 77, 3701 (1955).
287. Cohen, M. H. and Turnbull, D., Molecular transport in liquids and glasses, *J. Chem. Phys.*, 31, 1164 (1959).
288. Kovacs, A. J., Applicability of the free volume concept on relaxation phenomena in the glass transition range, *Rheol. Acta,* 5, 262 (1966).

289. Goldstein, M., Some thermodynamic aspects of the glass transition: free volume, entropy and enthalpy theories, *J. Chem. Phys.*, 39, 3369 (1963).
290. Kumar, S., The flow units in glasses, *Phys. Chem. Glasses,* 6, 147 (1965).
291. Macedo, P. B. and Litovitz, T. A., On the relative roles of free volume and activation energy on the viscosity of liquids, *J. Chem. Phys.,* 48, 245 (1965).
292. Mooney, M., A theory of the viscosity of a Maxwell elastic liquid, *Trans. Soc. Rheol.,* 1, 63 (1957).
293. Cohen, M. H. and Grest, G. S., Liquid-glass transition, a free-volume approach, *Phys. Rev. B,* 20, 1077 (1979).
294. Grest, G. S. and Cohen, M. H., Liquids, glasses, and the glass transition: A free-volume approach, *Adv. Chem. Phys.,* 48, 455 (1981).
295. Cohen, M. H. and Grest, G. S., A new free-volume theory of the glass transition, *Ann. N. Y. Acad. Sci.,* 371, 199 (1981).
296. Cohen, M. H. and Grest, G. S., The nature of the glass transition, *J. Non-Cryst. Solids,* 61/62, 749 (1984).
297. Adam, G. and Gibbs, J. H., On the temperature dependence of cooperative relaxation properties in glass-forming liquids, *J. Chem. Phys.,* 43, 139 (1965).
298. Mohanty, U., Cooperative relaxation in glass forming liquids, *Physika A,* 162, 362 (1990).
299. Chang, S. S., Bestul, A. B., and Horman, J. A., Critical configurational entropy of glass transition, in *Trav. VIIe Congress Intern. du Verre,* Bruxelles, 28 juin/3 juillet (1965), 26 1.13.
300. Greet, R. J. and Turnbull, D., Test of Adam-Gibbs liquid viscosity model with *o*-terphenyl specific-heat data, *J. Chem. Phys.,* 47, 2185 (1967).
301. Mohanty, U., On the nature of supercooled and glassy states of matter, *Physika A.,* 177, 345 (1991).
302. Richet, P., Viscosity and configurational entropy of silicate melts, *Geochim. Cosmochim. Acta,* 48, 471 (1984).
303. Hodge, I. M., Effects of annealing and prior history on enthalpy relaxation in glassy polymers. 6. Adam-Gibbs formulation of nonlinearity, *Macromolecules,* 20, 2897, (1987).
304. Hodge, I. M., Adam-Gibbs formulation of non-linear enthalpy relaxation, *J. Non-Cryst. Solids,* 131/133, 435 (1991).
305. Araujo, R. J., Entropy interpretation of viscosity of B_2O_3, *J. Chem. Phys.,* 44, 1299 (1966).
306. Scherer, G. M., Use of the Adam-Gibbs equation in the analysis of structural relaxation, *J. Am. Ceram. Soc.,* 67, 504 (1984).
307. Howell, F. S., Bose, R. A., Macedo, P. B., and Moynihan, C. T., Electrical relaxation in glass-forming molten salt, *J. Phys. Chem.,* 78, 639, (1974).
308. Tagliaferro, A., A thermodynamic approach to stretched exponential relaxation, *Mod. Phys. Lett. B,* 4, 1415 (1990).
309. Goldstein, M., On the temperature dependence of cooperative relaxation properties in glass-forming liquids. Comment on a paper by Adam and Gibbs, *J. Chem. Phys.,* 43, 1852 (1965).
310. Gupta, P. K., Negative pressure dependence of viscosity, *J. Am. Ceram. Soc.,* 70, C152 (1987).
311. Nemilov, S. V., Valence-configurational theory of the viscous flow of supercooled glass-forming liquids and its experimental validation, *Fiz. i Khim. Stekla (Sov. J. Glass Phys. Chem.),* 4, 129 (1978).

312. Nemilov, S. V., The nature of viscous flow of glasses with frozen structure and some consequences of the valence-configurational theory of fluidity, *Fiz. i Khim. Stekla (Sov. J. Glass Phys. Chem.),* 4, 662 (1978).
313. Douglas, R. W., The relation between the physical properties and the structure of glass. II. Viscosity as a function of temperature, *J. Soc. Glass Technol.,* 31, No. 141, 74 (1947).
314. Myuller, R. L., The nature of activation energy and experimental data on the flow behavior of refractory glass-forming materials, *Zh. Prikl. Khim. [J. Appl. Chem. (USSR)],* 28, 363 (1955).
315. Myuller, R. L., The valence theory of viscosity and flow behavior in the critical temperature range for refractory glass-forming materials, *Zh. Prikl. Khim. [J. Appl. Chem. (USSR)],* 28, 1077 (1955).
316. Smyth, H. T., Finlayson, J. R., and Remde, H. F., A theory of viscous flow of glasses, in *Trav. IVe Congrès Intern. du Verre,* Paris (1956), 317.
317. Eisenberg, A. and Teter, L. A., A relation between the relaxation mechanism and the distribution of relaxation times, *Polym. Lett.,* 7, 471, (1969).
318. Yinnon, H. and Cooper, A. R., Oxygen diffusion in multicomponent glass forming silicates, *Phys. Chem. Glasses,* 21, 204 (1980).
319. Farnan, I. and Stebbins, J. F., Observation of slow atomic motions close to the glass transition using 2-D ^{29}Si NMR, *J. Non-Cryst. Solids,* 124, 207 (1990).
320. Glasstone, S., Laidler, K. J., and Eyring, H., *Theory of Rate Processes,* McGraw-Hill, New York (1941).
321. Kobozev, N. I., On limitations of existing theories of chemical kinetics, *Zh. Fiz. Khim. (Russ. J. Phys. Chem.),* 28, 2067 (1954).
322. Faerber, G., Kim, S., and Eyring, H., The viscous flow and glass transition temperature of some hydrocarbones, *J. Phys. Chem.,* 74, 3510 (1970).
323. Levi, D. L., Note on the comparison of entropy and energy changes in diffusion and dielectric relaxation processes, *Trans. Faraday Soc.,* 42A, 152 (1946).
324. Eley, D. D., The kinetics of haemoglobin reactions, *Trans. Faraday Soc.,* 39, 172 (1943).
325. Hanggi, P., Talkner, P., and Borkovec, M., Reaction-rate theory: Fifty years after Kramers, *Rev. Mod. Phys.,* 62, 251 (1990).
326. Hinshelwood, C. N., On the theory of unimolecular reactions, *Proc. R. Soc. London,* 113A, 230 (1927).
327. Polanyi, M. and Wigner, E., Über die Interferenz von Eigenschwingungen als Ursache von Energieschwankungen und chemischer Umsetzungen, *Z. Phys. Chem.* 139A, 439 (1928).
328. Moelwyn-Hughes, E. A., *Chemical Statics and Kinetics of Solution,* Academic Press, New York-London, (1972), 282, 306.
329. Dushman, S., A theory of chemical reactivity. Calculation of rates of reactions and equilibrium constants, *J. Am. Chem. Soc.,* 43, 397 (1921).
330. Nemilov, S. V., The kinetics of elementary processes in the condensed state. IV. Activation entropy of corpuscular transport processes, *Zh. Fiz. Khim. (Russ. J. Phys. Chem.),* 43, 1433 (1969).
331. Nemilov, S. V., The kinetics of elementary processes in the condensed state. II. Shear relaxation and the equation of state for solids, *Zh. Fiz. Khim. (Russ. J. Phys. Chem.),* 42, 1391 (1968).
332. Frenkel', Ya. I., *Kinetic Theory of Liquids,* in *Sobranie Izbrannych Trudov* (Collected Works), Izdatel'stvo Akademii Nauk SSSR, Moscow-Leningrad (1959), v. 3, 206.

333. Nemilov, S. V., Relationship between the velocity of sound, mass, and energy of chemical interaction, *Dokl. Akad. Nauk SSSR* (Proc. Acad. Sci. USSR), 181, 1427 (1968).
334. Tobolsky, A., Powell, R. E., and Eyring, H., Elasto-viscous properties of matter, in *Frontiers in Chemistry*, Burk, R. E. and Grummit, O., Eds., Interscience Publ. Inc., New York (1943), v. 1, 125.
335. Berry, G. C. and Fox, T. G., The viscosity of polymers and their concentrated solutions, *Adv. Polymer Sci.*, 5, 261 (1968).
336. Nemilov, S. V., The relation between the free energy of activation for viscous flow and the energy of chemical bonds in glasses, *Sov. Phys. - Solid State*, 6, 1075 (1964).
337. Nemilov, S. V., The energetics and properties of glassy and crystalline solids, in *Stekloobraznoje Sostojaniye. Trudy 5 Vsesoyuznogo Soveshaniya po stekloobraznomu Sostojaniyu* (*The Vitreous State*. Proc. 5th All-Union Conf. on the Glassy State, Leningrad, May 26-30, 1969), Porai-Koshitz, E. A., Ed., Nauka, Leningrad (1971), 10.
338. Myuller, R. L., Chemical characteristics of polymeric glass-forming substances and the nature of glass formation, in *The Structure of Glass*, (Proc. 3rd All-Union Conf. on the Glassy State, Leningrad, November 16-20, 1959), v. 2, Porai-Koshitz, E. A., Ed., (engl. transl. Uvarov, B., Ed.), Consultants Bureau, New York (1960), 50.
339. Takahashi, K. and Osaka, A., Bond dissociation energies in the alkali disilicate, digermanate and metaphosphate glasses, *J. Ceram. Soc. Jpn.*, 91, 389 (1983).
340. Steblevsky, A. V., Alichanjan, A. S., Pashinkin, A. S., and Gorgoraki, V. I., Formation heats of arsenic sulphides, *Zh. Neorg. Khim. (Russ. J. Inorg. Chem.)*, 31, 1650 (1986).
341. Steblevsky, A. V., Alichanjan, A. S., Gorgoraki, V. I., and Pashinkin, A. S., Vaporization processes in the system As-Se, *Zh. Neorg. Khim. (Russ. J. Inorg. Chem.)*, 31, 834 (1986).
342. Boone, S. and Kleppa, O. J., Enthalpies of formation for group IV selenides ($GeSe_2$, $GeSe_2$(am), SnSe, $SnSe_2$, PbSe) by direct-combination drop-calorimetry, *Thermochim. Acta*, 197, 109 (1992).
343. Karapet'yants, M. Ch. and Karapet'yants, M. A., Eds., *Basic Thermodynamic Constants of Inorganic and Organic Substances*, Khimiya, Moscow (1968).
344. Gurvich, L. V., Karachevtsev, G. V., Kondrat'ev, V. N., Lebedev, Yu. A., Medvedev, V. A., Potapov, V. K., and Chodeev, Yu. S., *Energies of Chemical Bonds Breaking. Ionization Potentials and Affinity to Electron*, Nauka, Moscow (1974).
345. Glushko, V. P., Ed., *Thermodynamic Characteristics of Individual Substances* Nauka, v. I-IV, Moscow (1978-1982).
346. Dodd, Ch. G. and Glen, G. L., Studies of chemical bonding in glasses by X-ray emission spectroscopy, *J. Am. Ceram. Soc.*, 53, 322 (1970).
347. Pauling, L., The nature of silicon-oxygen bonds, *Am. Mineralogist*, 65, 321 (1980).
348. Lesikar, A. V., On the self-association of the normal alcohols and the glass transition in alcohol-alcohol solutions, *J. Solution Chem.*, 6, 81 (1977).
349. Donth, E.-J., *Glasübergang*, Akademie-Verlag, Berlin, (1981), 168.
350. Nemilov, S. V., Relationship between the configurational entropy and activation entropy of viscous flow of supercooled glass-forming liquids, *Fiz. i Khim. Stekla (Sov. J. Glass Phys. Chem.)*, 2, 193 (1976).

351. Angell, C. A., Two-state thermodynamics and transport properties for water as zeroth-order results of a "Bond Lattice" model, *J. Phys. Chem.,* 75, 3698 (1971).
352. Filipovich, V. N. and Kalinina, A. M., On the nature and interrelation of changes in glass properties upon vitrification, in *Stekloobraznoje Sostojaniye,* Trudy 5 Vsesoyuznogo Soveshaniya po Stekloobraznomu Sostoyaniu (Proc. 5th All-Union Conf. on the Glassy State, Leningrad, May 26-30, 1969), Porai-Koshitz, E. A., Ed., Nauka, Leningrad (1971), 28.
353. Nemilov, S. V., The viscosity and elastic properties of As-S melts and glasses and their valence structure, *Fiz. i Khim. Stekla (Sov. J. Glass Phys. Chem.),* 5, 398 (1979).
354. Napolitano, A., Simmons, J. H., Blackburn, D. H., and Chidester, R. E., Analysis of low temperature viscosity data for three NBS standard glasses, *J. Res. Nat. Bur. Stand.,* 78A, 323 (1974).
355. Sasabe, H., DeBolt, M. A., Macedo, P. B., and Moynihan, C. T., Structural relaxation in an alkali-lime-silicate glass, in *Proc. XI Intern. Congr. Glass,* Prague (1977), 339.
356. Greet, R. J. and Turnbull, D., Glass transition in *o*-terphenyl, *J. Chem. Phys.,* 46, 1243 (1967).
357. Cukierman, M., Lane, J. W., and Uhlmann, D. R., High-temperature flow behavior of glass-forming liquids: a free-volume interpretation, *J. Chem. Phys.,* 59, 3639 (1973).
358. Moynihan, C. T., Mass transport in fused salts, in *Ionic Interactions from Dilute Solutions to Fused Salts,* Petrucci, S., Ed., v. I, Acad. Press, New York (1971), 261.
359. Nemilov, S. V., The kinetics of elementary processes in the condensed state. VII. The activation volume of corpuscular transport processes and its temperature dependence; viscous flow of glasses, *Zh. Fiz. Khim. (Russ. J. Phys. Chem.),* 45, 2747 (1971).
360. Mazurin O. V. and Stolyar, S. V., Secondary structural relaxation and its probable effect on the viscosity of sodium borosilicate glasses, *Fiz. i Khim. Stekla (Sov. J. Glass Phys. Chem.),* 15, 480 (1989).
361. Sizov, S. N., Empirical relation between the parameters of the temperature dependence of viscosity for glasses with frozen and equilibrium structures, in *Vsesoyuznyi Simpozium po Relaksationnym Yavleniyam v Neorganicheskich Steklach, Tesisy Dokladov* (All-Union Symposium on Relaxation Processes in Inorganic Glasses, Rep. Abstracts), Tbilisi (1984), 49.
362. Wu, L. and Nagel, S. R., Secondary relaxation in *o*-terphenyl glass, *Phys. Rev. B,* 46, 11198 (1992).
363. Weber, T. A. and Stillinger, F. H., Tiling model for glass formation with incremental domain-size kinetics, *Phys. Rev. B,* 36, 7043 (1987).
364. Nechoda, A. R., Rostiashvili, V. G., Irzak, V. I., and Irzak, T. F., Influence of temperature prehistory on α-β transitions in polymers, *Vysokomol. Soed., Kratkye Soobshch. [Polym. Sci. (USSR), Short rep.],* 30, 768 (1988).
365. Shlesinger, M. F., Fractal time in condensed matter, *Ann. Rev. Phys. Chem.,* 39, 269 (1988).
366. Bendler, J. T. and Shlesinger, M. F., Generalized Vogel Law for glassforming liquids, *J. Statist. Phys.,* 53, 531 (1988).
367. Gupta, P. K., A gradient energy formalism for the nonexponential structural relaxation kinetics, *J. Non-Cryst. Solids,* 102, 250 (1988).

368. Torell, L. M., Börjesson, L., and Elmorth, M., On the non-exponential versus non-Arrhenius behaviour of the structural relaxation in glass-forming liquids, *J. Phys.: Cond. Matter,* 2, Suppl. A, SA207 (1990).
369. Flesseles, J.-M. and Botet, R., Derivation of a stretched-exponential time relaxation, *J. Phys. A,* 22, 903 (1989).
370. Nagel, S. R. and Dixon, P. R., Relation between stretched exponential relaxation and Vogel-Fulcher behaviour above the glass transition, *J. Chem. Phys.,* 90, 3885 (1989).
371. Tagliaferro, A., A thermodynamic approach to stretched exponential relaxation, *Mod. Phys. Lett., B,* 4, 1415 (1990).
372. Curant, Ch., A thermodynamic theory of relaxation based on a distribution of non-linear processes, *J. Non-Cryst. Solids,* 131/133, 196 (1991).
373. Williams, G., Molecular motion in glass-forming liquids, *J. Non-Cryst. Solids,* 131/132, 1 (1991).
374. Popoular, M., On slow relaxations in fragile glasses, *Philos. Mag. Lett.,* 64, 421 (1991).
375. Etienne, S., Cavaille, J. Y., and Perez, J., Molecular mobility in the liquid-glass transition range: cooperative motions, *J. Non-Cryst. Solids,* 131/133, 66 (1991).
376. Mansfield, M. L., Volume relaxation of glasses, *J. Chem. Phys.,* 94, 7521 (1991).
377. Ngai, K. L. and Rendell, R. W., Fundamental issues confronting models of non-linear structure relaxation, in *Physics of Non-Crystalline Solids,* Pye, L. D., LaCourse, W. C., and Stevens, H. J., Eds., Taylor & Francis, London-Washington (1992), 309.
378. Fredrickson, G. H., Linear and nonlinear experiments for a spin model with cooperative dynamics, *Ann. N.Y. Acad. Sci.,* 484, 185 (1986).
379. Butler, E. and Harrowell, P., The origin of glassy dynamics in the 2D facilitated kinetic Ising model, *J. Chem. Phys.,* 95, 4454 (1991).
380. Götze, W., *Liquid-Glass Transition,* Neuchâtel: University Press (1987).
381. Hiwatari, Y. and Miyagawa, H., Dynamical singularities near the liquid-glass transition: Theory and molecular dynamics study, *Solid State Ionics,* 47, 179 (1991).
382. Li, G., Du, W. M., Chen, X. K., Cummins, H. Z., and Tao, N. J., Testing mode-coupling predictions for α- and β-relaxation in $Ca_{0.4}K_{0.6}(NO_3)_{1.4}$ near the liquid-glass transition by light scattering, *Phys. Rev. A,* 45, 3867 (1992).
383. Nemilov, S. V., On the definition of the concept "Vitreous State," *Fiz. i Khim. Stekla (Sov. J. Glass Phys. Chem.),* 17, 511 (1991).
384. Van Santen, R. A., The Ostwald rule, *J. Phys. Chem.,* 88, 5768 (1984).
385. Hallbruckner, A., Mayer, E., and Johari, G. P., Glass transition in pressure-amorphized hexagonal ice. A comparison with amorphous forms made from the vapor and liquid, *J. Phys. Chem.,* 93, 7751 (1989).
386. Hazen, R. M., Figner, L. W., Hemley, R. J., and Mao, H. K., High-pressure crystal chemistry and amorphysation of α-quartz, *Solid State Commun.,* 72, 507 (1989).
387. Winters, R. R., Garg, A., and Hammack, W. S., High-resolution microscopy of pressure-amorphized α-quartz, *Phys. Rev. Lett.,* 69, 3751 (1992).
388. Tagantsev, D. K. and Nemilov, S. V., Photoviscosity effect in chalcogenide glasses, *Fiz. i Khim. Stekla (Sov. J. Glass Phys. Chem.),* 15, 397 (1989).
389. Utsugi, Y. and Mizuchima, Y., Lattice-dynamical aspects in photoexcited chalcogenide glasses, *Jpn. J. Appl. Phys.,* 31, 3922 (1992).

Index

C

D

G

H

I

J

K

L

M

N

O

P

Q

R

S

T

U

V

W

X

Z